Bioreactors

Animal Cell Culture Control
for Bioprocess Engineering

Bioreactors
Animal Cell Culture Control for Bioprocess Engineering

Goutam Saha
Alok Barua
Satyabroto Sinha

CRC Press
Taylor & Francis Group
Boca Raton London New York

CRC Press is an imprint of the
Taylor & Francis Group, an **informa** business

First published in paperback 2018

First published 2015 by
CRC Press
Taylor & Francis Group
6000 Broken Sound Parkway NW, Suite 300
Boca Raton, FL 33487-2742

International Standard Book Number-13: 978-1-4987-3599-5 (Hardback)
International Standard Book Number-13: 978-1-138-74968-9 (Paperback)

Visit the Taylor & Francis Web site at
http://www.taylorandfrancis.com

and the CRC Press Web site at
http://www.crcpress.com

Contents

Preface

Bioprocess engineering is the most important component for industries that produce commercial products such as industrial alcohol, organic solvent, and baker's yeast, special products such as antibiotics, antibodies, therapeutic proteins, vaccines, and recombinant products such as insulin. The bioreactor is where the bioprocess operation takes place, and its design and controlled operation are important for production aspects such as purity, quantity, efficiency, and safety.

Several excellent textbooks on bioprocess engineering are currently available; these generally are written almost exclusively with either biotechnology or chemical engineering in mind. However, these books do not cover the details of control system engineering. *Bioreactors: Animal Cell Culture Control for Bioprocess Engineering* will bridge the gap between control system engineering and biotechnology. The rigorous theoretical analysis it presents for the control of nonlinear systems such as bioprocesses is the major strength of the book.

Bioprocesses require effective control techniques due to increased demand on productivity, product quality, and environmental responsibility. This is especially important where the biomaterials are costly and require stringent control over product formation, as in animal cell cultures. Animal cell culture technologies are used for the production of many enzymes, hormones, vaccines, monoclonal antibodies, and anticancer agents. Among them, the most important product is monoclonal antibodies, which are produced by hybridoma cells.

Monoclonal antibodies have been used as diagnostic agents to develop many drugs, toxins, vitamins, and other biological compounds.

There are several types of bioreactors used in the laboratory as well as in large-scale industrial applications. Some of these are continuous stirred tank, bubble column, airlift, see-saw, and packed bed reactors. As their names indicate, they are meant for aerobic bioreaction processes. The technique used for fermentation of animal cells differs from that used with bacteria, yeasts, and fungi. The typical size of animal cells is 10–30 μm. Animal cells do not have a cell wall, but are surrounded by a thin and fragile plasma membrane, and therefore the cells are very sensitive to shear force. For the efficient growth of cells, the fermenter or bioreactor should operate with essentially no or minimum shear force.

The design, fabrication, and control of a new type of bioreactor meant especially for animal cell culture are covered in this book. Existing bioreactors like the continuous stirred tank reactor, bubble column reactor, and airlift reactors are briefly discussed. However, these conventional reactors are not suitable for fermentation of animal cells, as they cause shear damage to the cells. Moreover, large support material surface areas are to be provided for anchorage of dependent cells. The laboratory-scale cultivation of animal cells is carried out in T-flasks, spinner flasks, and other extremely small reactors, and these are not suitable for any commercial production.

The new bioreactor is called the see-saw bioreactor and is ideal for the growth of cells with a sensitive membrane. The name is derived from its principle of operation, in which liquid columns in either limb of the reactor alternately go up and down. The oxygen transfer in this type of bioreactor has been studied by a distributed parameter model. The working volume of the reactor is small, to within 15 L. However, it can be easily scaled up for large production in volume of cell mass in the drug and pharmaceutical industries. The see-saw bioreactor was developed at the Indian Institute of Technology Kharagpur. The authors neither experimented on animal cells nor made any animal cell culture in the prototype reactor developed and patented due to limited facilities in the department; however, its description and analysis clearly command its suitability for the said culture.

The primary aim of this book is to describe the principle of operation of a new type of bioreactor and how to automatically control the bioprocess. In this context different control strategies have been discussed. The authors have conducted thorough experimental research on this prototype bioreactor and applied a time delay control for yield maximization. Emphasis has been placed on the development of a suitable control strategy and mode of operation such that more products can be made from this bioreactor. However, the model that is developed from the mass balance concept does not describe the bioprocess completely. The model parameters often vary with time due to metabolic variations and physiological and genetic modifications. The reproducibility of the biotechnological experiments is also poor. Thus a bioprocess, in general, is a nonlinear, undermodeled multivariable system with uncertainties.

The limitations of conventional control such as proportional-integral-derivative have been discussed here. Time delay control, which has never before been used as a controller in bioreactor control, has been applied. It is designed as a tracking controller such that the process variables allow optimal trajectories within finite error bounds. In a bioprocess some of the process variables are not measurable. A suitable observer, or "software sensor," has been designed. The optimal trajectories of different bioprocess variables have been derived using genetic algorithms. However, separate controllers are used for controlling the temperature and pH of the bioreactor fluid.

The authors would like to express their appreciation and gratitude to the many individuals who have contributed to the development of the see-saw bioreactor. Our best wishes go to the students and researchers who are the ultimate users of this book.

Goutam Saha
Alok Barua
Satyabroto Sinha
Kharagpur

MATLAB® is a registered trademark of The MathWorks, Inc. For product information, please contact:

The MathWorks, Inc.
3 Apple Hill Drive
Natick, MA 01760-2098 USA
Tel: 508-647-7000
Fax: 508-647-7001
E-mail: info@mathworks.com
Web: www.mathworks.com

Authors

Goutam Saha earned his BE in electrical engineering, his ME in electronics and telecommunication engineering, and his PhD in electrical engineering from Bengal Engineering and Science University (formerly known as B.E. College) and IIT Kharagpur in 1984, 1989, and 1999, respectively. He worked as a postdoctoral research fellow at Ben Gurion University, Israel, from January to August 2002. He is currently a full professor in the Department of Information Technology, North-Eastern Hill University, Shillong, India. With more than 25 years of teaching experience in various engineering colleges including the National Institute of Technology (Durgapur), the National Institute of Technical Teachers Training and Research (Kolkata), and other government engineering colleges of West Bengal. He has published several papers in the areas of instrumentation, bioreactor design and control, bioinformatics, and system biology. He has supervised many MS theses and five PhD theses on bioinformatics and system biology. He is also one of the team members holding a patent for the design of a see-saw bioreactor. He is a member of the Sixth Reconstituted Task Force Committee on Bioinformatics, Computational and System Biology for the Government of India.

Alok Barua received his bachelor of technology in instrumentation and electronics engineering, master of electronics and telecommunication engineering, and PhD in electrical engineering from Jadavpur University and Indian Institute of Technology (IIT) Kharagpur in 1977, 1980, and 1992, respectively. He is full professor in the Department of Electrical Engineering, IIT Kharagpur. With more than 30 years of teaching experience at IIT, he has published many papers in his teaching and research areas: instrumentation, bioreactor design and control, testing and fault diagnosis of analog and mixed-signal circuits, and image processing. He has supervised several MS theses and two PhD theses on bioreactor control and instrumentation. He also holds a patent for the design of the see-saw bioreactor. He has delivered invited lectures at many different universities in the United States, Europe, and the Far East. He has worked as a visiting professor, guest professor, and research professor at the University of Arkansas, Fayetteville; University of Karlsruhe and Frankfurt University (Germany); Yonsei University and Korea University (Seoul); and other institutions. Dr. Barua is a senior member of the Institute of Electrical and Electronics Engineers.

Professor Barua is the author or coauthor of several books: *Computer Aided Analysis, Synthesis and Expertise of Active Filters* (Dhanpat Rai, New Delhi, 1995), *Fault Diagnosis of Analog Integrated Circuit* (Springer, the Netherlands, 2005), *Fundamentals of Industrial Instrumentation* (Wiley India, New Delhi, 2011), and *Analog Signal Processing: Analysis and Synthesis* (Wiley India, New Delhi, 2014). He also coauthored a research monogram, *3D Reconstruction with Feature Level Fusion* (Lambert Academic, Germany, 2010).

Satyabroto Sinha (deceased) earned a PhD in electrical engineering from the Indian Institute of Technology (IIT) Kharagpur. He taught electrical engineering at IIT Kharagpur from December 1963 to May 2005 (professor, 1982–2002; professor emeritus, 2002–2005). He advised 13 doctoral students, coauthored several books, and published approximately 60 conference and journal papers, nationally and internationally. His area of specialization was instrumentation and control. He was a national coordinator for the Technology Development Mission on Communication, Networking and Intelligent Automation for the Government of India from 1995 to 2000. He was a senior member of the Institute of Electrical and Electronics Engineers and a chartered engineer and fellow of the Institution of Engineers (India).

CHAPTER ONE

Introduction

1.1 A new type of bioreactor

Nowadays, to cope with various diseases—new or old—in terms of vaccinations and an improved variety of drug production, we have to culture animal cell lines. The main difficulty with culturing animal cell lines is that the cell membrane of animal cells is very thin and weak, so many difficulties crop up when culturing animal cell lines with the existing conventional bioreactors. For example, in the case of the continuous stirred tank reactor (CSTR), a substantial amount of animal cells will be destroyed by the impinging fan blades and the resultant shear force generated inside the bioreactor. Many cells may also be destroyed because of entrapment in the air bubbles meant for aeration. These are also valid issues in the case of bubble column or airlift type of bioreactors. This book describes the design and development of a new type of bioreactor suitable for animal cell line culture.

In this bioreactor the above-mentioned difficulties, which cause cell death, are absent due to this somewhat different design. This novel bioreactor is called the *see-saw bioreactor* [1]. The name was derived from its underlying principle of operation. The working volume of the prototype bioreactor is small (within 15 L), but we cannot underestimate its importance, as the cost of the enzymes (and so forth) produced from this animal cell line culture is very high. Moreover, the cost of the substrate used is also very high; so it is not profitable to use commercial size bioreactors for this purpose. The above justifies the cost-effectiveness of the design and development of a smaller reactor for animal cell line culture. Rather, emphasis should be given to developing a suitable control strategy so that more and more enzymes and metabolites can be produced from these small-volume bioreactors. These are the objectives we seek to fulfill in the research study represented in this book.

1.2   Bioreactor modeling and control

A fermentation process can be explained as follows: "the microorganisms (bacteria, fungi, yeast, etc.) grow with the consumption of certain nutrients (carbon derivatives, N, P, K, etc.) assuming that the environmental conditions are favourable" [2]. The various objectives of fermentation are [3]

1. To produce biomass (e.g., baker's yeast production)
2. To extract products of interest from the biomass, secondary metabolites, and so forth (antibiotics, ethanol, different enzymes, etc.)
3. To control pollution (aerobic and anaerobic digestion of different carbon substrates by suitable microorganisms of interest)

Designing a suitable bioprocess controller is to achieve maximization of the product. This is not an easy task because of the following reasons [4]:

1. Biochemical processes exhibit nonlinear dynamic behavior, usually represented by nonlinear differential equations.
2. Biochemical processes involve living organisms, so their dynamic behavior is not only nonlinear, but also of the time-varying parameter (TVP) type. This happens because of complex biological phenomena involving genetic mutation, metabolic variations of cell mass resulting from unknown physiological variations, and so forth.
3. Biochemical processes are multi-input and multi-output processes with constraints.
4. Until today most bioprocesses have been poorly understood, so the models developed have represented them poorly. Many unknown dynamics cannot be incorporated in the models.
5. Further, most of the process variables are not directly measurable, as suitable transducers for the purpose are not available. So even if efforts are made to model the bioprocess in

detail by taking into account more variables, many of them will remain unmeasurable. This will pose a problem to controller development using the detailed model.

6. There are various factors that can influence a bioprocess at any instant in time. Reasons for most of these unexpected disturbances are unknown to us. This also explains why reproducibility of biochemical experiments is poor.

The normal strategy for designing a controller for such a system involves linearizing the nonlinear process around a steady-state operating point and then applying the standard linear control theory to the small-signal model. This type of controller may produce good results if the process is confined to the neighborhood of the operating point. But for highly nonlinear processes like the bioprocess, and where the operating point is continually changing, this type of linearized controller produces poor results. It generally fails in scale-up conditions. For example, multiloop control [5] and adaptive control [6] do not produce expected results. The same is also true for fuzzy [7–10] and neurofuzzy controllers [11]. Although fuzzy controllers do not use standard process models [12], the results are still not encouraging because of the poor reproducibility of the bioprocesses.

In this book, an effort has been made to develop suitable algorithms to reconstruct the unmeasurable states. This is termed software sensors [13–15]. Then a controller has been designed that takes care of the nonlinearities and nonstationarities of the parameters and the undermodeled dynamics of the bioprocess. The controller forces the states to track respective predefined optimal time trajectories by generating suitable control action. The optimal time trajectories are generated using genetic algorithms [16].

To start with, a generalized bioprocess model having five states and four inputs has been formulated [17]. There are nine operating modes in which a bioprocess can be operated. First, an operating mode is selected for a particular process, and then genetic algorithms are applied to find optimal time trajectories of the states. The control strategy follows suit. This will maximize production.

1.3 Organization of the book

The book is organized as follows. In addition to this introductory chapter, we have the following chapters:

Chapter 2   This chapter describes the design and development of a novel see-saw bioreactor. Efforts have been made to derive an expression for the mass transfer coefficient of gaseous oxygen to a liquid medium of the bioreactor with certain assumptions. Experimentation has been carried out to validate the expression. The results are presented.

Chapter 3   This chapter presents the basic generalized mathematical model (unstructured and unsegregated) to describe the kinetics of fermentation. The nine possible operating modes of the bioreactor include the conventional batch, continuous, and fed-batch operating modes. An expert system package (BIPROSIM) has been developed that helps in selecting the best operating mode for a particular fermentation process for given initial conditions. It is possible to use this package for real-time simulation. Suitable parameter estimation algorithms are required to update the TVPs and also predict the best operating mode for the sampling interval in Chapter 4.

Chapter 4   This chapter presents the genetic algorithms used for deriving the optimal time trajectories of the bioprocess variables and the corresponding control inputs for maximizing the product yield. The bioprocess model discussed in Chapter 3 has been used. After selecting the best operating mode (op-mode), we apply genetic algorithms such that the product concentration is maximized. Thus optimal time profiles of the process variables are generated. These time profiles of the states will act as reference trajectories for later control. The optimal time trajectories of the states for single cell protein (SCP) production in all nine op-modes have been displayed.

Chapter 5   This chapter dwells on the application of time delay control to bioprocesses. The controller maneuvers the control inputs in such a way that the process variables such as biomass concentration, substrate concentration, product concentration, and oxygen concentration follow the desired time trajectories.

Since the dissolved oxygen concentration is the only measurable variable, a suitable observer is required to estimate cell mass concentration, substrate concentration, and product concentration. The observer acts as a "software sensor." The different inputs are the feed rates, recycling, peristaltic pump infusion, and oscillation time period of the bioreactor liquid, which in turn is proportional to the mass transfer coefficient of gaseous oxygen to the bioreactor liquid. Simulation results for different operating modes have been presented in this chapter.

Chapter 6   This chapter presents the working setup in the laboratory of the bioreactor. The instrumentation system for automated operation of the fabricated prototype see-saw bioreactor is described. To verify time delay control (TDC), the biomass production of yeast is investigated. These steps are followed:

1. Formulation of the model equations for the biomass growth of yeast
2. Selection of the operating mode and application of the genetic algorithm to derive the optimal time profiles of the bioprocess variables
3. Application of TDC for following the optimal time profiles in the laboratory
4. Off-line analysis of the bioprocess yield
5. Comparison of experimental and simulated results
6. Discussion of the results

Chapter 7   This chapter sums up the investigation. It further defines the future scope of work in terms of design, control, and software sensors.

Appendix   The appendix presents the development of environmental controllers (temperature and pH controllers). For controlling the temperature and pH of the bioreactor medium to within a band, on–off and proportional controllers, respectively, are used. They become operative as soon as the process starts working and remain operative until the end. The algorithms for temperature and pH control are also given.

CHAPTER TWO

Novel see-saw bioreactor

A see-saw bioreactor has been developed for the cultivation of animal cells, which are fastidious to culture and very sensitive to shear effects. The purpose of this study is to improve the understanding of the interactions among various parameters that govern the oxygen transfer phenomenon in this type of bioreactor. A distributed parameter model for gaseous oxygen transfer to the liquid phase (substrate) has been derived. The proposed model predictions are compared with experimental results.

Compared to other widely used cell lines for various purposes (e.g., antibiotic production and enzyme production), animal cell line culture is more difficult. This is because animal cell membranes are very weak and sensitive to shear effects. Conventional bioreactors like the continuous stirred tank reactor (CSTR), bubble column bioreactor, and airlift bioreactors are not very suitable for this purpose. In the case of CSTR, many of the cells are destroyed by impinging fan blades of the stirrer. In the case of the bubble column bioreactor, cell death occurs by entrapment of the cells in the bubbles and during the bubble rupture. Various bioreactors have been developed [18,19] by researchers for animal cell culture.

A see-saw bioreactor uses no mechanical accessories for dissolving oxygen. Since there is no mechanical agitation system, only a little shear force is generated. This particular feature makes this bioreactor very attractive for high-density animal cell line culture. The oxygen transfer enhancement is through periodic renewal of the exposed liquid surface, and hence higher productivity can be achieved by using even a small laboratory-scale bioreactor. This underscores the potential of this simple bioreactor configuration.

2.1 Construction and working of the see-saw bioreactor

The see-saw bioreactor is a 15 L aerobic bioreactor. In the aerobic process, oxygen transfer to the reaction phase (liquid) is an important consideration. The transfer of oxygen to an aerobic culture and the transfer of carbon dioxide from the culture to the exhaust are the two most important mass transfer considerations in the process.

Oxygen transfer from the gas phase to the liquid phase has been the subject of much research [20,21]. The main problem of oxygen transfer to a bioreactor system is the poor solubility of oxygen in water. The equilibrium concentration is usually estimated by applying Henry's law as the phase equilibrium relationship:

$$p_o = H_o C_o \tag{2.1}$$

where

p_o = partial pressure of oxygen
H_o = Henry's law of constants for oxygen
C_o = concentration of oxygen in liquid

Improvement in oxygen transfer can be achieved by increasing turbulence, interfacial area, and partial pressure of oxygen in the system and renewal of the surface for mass transfer of oxygen. However, increasing the partial pressure of oxygen to enhance the mass transfer rate of oxygen adversely affects the rate of carbon dioxide release. Packed beds, bubble column reactors, and so on, provide a high interfacial area. High-pressure drops and possibilities of cell damage are often the limiting conditions in such equipment. Increasing turbulence and mixing provide a higher mass transfer rate. But in animal cell culture operations, cell damage at high turbulence is a serious problem. In addition, breaking of substrate particle may limit the degree of turbulence. Therefore, the functional requirements of an efficient aerobic cell culture bioreactor system are summed up as follows:

1. High transfer rate with minimum turbulence leading to minimum damage to the cell and substrate material
2. Ease of operation
3. Ease of construction
4. Compactness

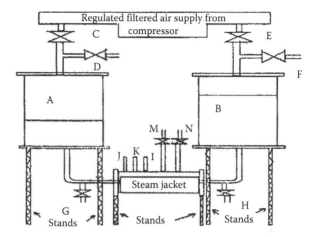

FIGURE 2.1   Schematic block diagram of the see-saw bioreactor.

The mass transfer of oxygen in the see-saw effect bioreactor under study is based on renewal of the surface of mass transfer without much bulk fluid turbulence. As shown in Figure 2.1, this bioreactor has two identical cylinders with uniform cross sections (marked A and B). These two cylinders are connected by a flow pipe and sensor assembly system. The opening of each cylinder is connected to two normally closed solenoid valve assembly. C and D are connected to one cylinder, and E and F are connected to the other cylinder. Solenoid valves C and E are connected to an air compressor through an air filter and pressure regulating system. D and F are vented to the atmosphere (through air filters). The aim of this assembly is to transfer liquid from one column to another for a stipulated period. The flow is reversed at the end of the period, for the same amount of time. This constitutes one cycle of operation. This is achieved as follows:

1. An "astable multivibrator" with on–off time adjustment is used. The same can be activated using a computer program.

2. During the ON time solenoid valves C and F are open. During this period valves E and D remain closed. During the OFF period E and D are open and valves C and F are closed.

3. During the ON time air from the compressor pushes the liquid in column A toward B, and trapped air in B finds its way out through valve F. As a result the liquid level

in column A falls and the liquid level in column B rises equally, with cylinders being similar. This is continued for a predetermined period.

4. During the OFF period solenoid valves C and F are de-energized and valves E and D are energized. That is, air at high pressure from the compressor enters column B through valve E, and the liquid column in B is pushed downward. At the same time the liquid column in A rises upward and the entrapped air in A is vented through the valve D. This is maintained for the same predefined time period.

5. Oscillation of the liquid column is obtained in the experimental bioreactor by repeating steps 3 and 4.

Mass transfer of gaseous oxygen to the liquid medium takes place

1. Through the falling liquid film on the surface of the vessel wall. The mass transfer rate of oxygen is a function of hydrodynamic conditions in the film.

2. Through the flat top surface of the liquid in each cylinder.

The oxygen accumulated in the falling film during the receding of the liquid level in one arm is deposited as dissolved oxygen in the bulk liquid. The turbulence of the system or the oscillation is visible and is very mild for all practical purposes. As shown in Figure 2.1, sensors attached to the bioreactor system are temperature sensor j (Platinum Resistance Temperature Detector (RTD): PT-100 type), pH sensor k, and dissolved oxygen sensor I. G and N are the ports where pH balancing peristaltic pumps have been connected. M and H are the ports where feed-in and feed-out pumps have been connected.

2.2 Theoretical modeling and simulation

Theoretical study of the oxygen transfer rate in this setup consists of two parts:

1. Modeling and study of oxygen transfer in the falling film

2. Modeling and study of oxygen transfer across the flat liquid surface

These are added up to provide the total oxygen transfer in the equipment.

Oxygen transfer to the falling film

The analysis is based on the following assumptions:

1. The fluid (liquid) is Newtonian in nature and incompressible.

2. Variation of viscosity is independent of position and constant throughout the film. Also, it does not differ substantially during an experimental run.

3. The end and entrance effects are negligible.

4. The change of momentum in the direction of the thickness of the film is small and therefore neglected.

The mass transfer through falling film has been studied by many workers [22]. The mass transfer analysis is based on the following further assumptions:

1. The falling film is laminar.

2. The diffusion takes place slowly in the liquid film so that the penetration distance is small in comparison to the film thickness.

3. The fluid properties are assumed to be constant—invariant with respect to location and time.

4. The end effects are neglected; that is, the film is long.

5. The interface solute concentration in the liquid is taken to be the solubility of gas in liquid.

6. Diffusion in the vertical direction is neglected compared to convective effects.

The mass flux when a solute from a gas phase is transferred to a stagnant liquid pool is a function of diffusivity, time of exposure, and the concentration gradient. Based on this, the mass flux is given as

$$N_a = (Cai - Ca0)\sqrt{\frac{4D_{ab}}{\pi t_{\exp}}} \qquad (2.2)$$

where

N_a = mass transfer flux (g/cm^2 s)
D_{ab} = diffusivity of A into B (cm^2/s)
Cai = interface oxygen concentration (g/cm^3)
$Ca0$ = bulk oxygen concentration (g/cm^3)
$t_{\exp}$ = time of exposure (s)

Let

$$K_L = \sqrt{\frac{4D_{ab}}{\pi t_{exp}}}$$

So, the equipment mass transfer of oxygen through the whole area of the liquid film of the bioreactor for exposure time t_{exp} is given by the following expression:

$$G_\omega = K_L a(Cai - Ca0)t_{exp}$$

where

a = total area of liquid film on the wall of the bioreactor in cm^2

From Figure 2.2, the exposure time of the dz element in one cycle is calculated to be

$$t_{exp}|dz = 2\left(t_C - \frac{Z}{u_S}\right)$$

where

t_C = time period of one half of the cycle

Z = height of the location of the dz element on the falling film from the top of the liquid film (reference level)

u_S = surface velocity of the falling film

Oxygen transfer to the dz element of the falling liquid film of a limb of the bioreactor in one cycle is given as

$$dGw = 2\pi r dz \left(Cai - Ca0\right) 2\left(t_C - \frac{Z}{u_S}\right) \sqrt{\frac{4\,D_{ab}}{\pi 2\left(t_C - \frac{Z}{u_S}\right)}}$$

or

$$dGw = 4\sqrt{2}\pi r(Cai - Ca0)\sqrt{\frac{D_{ab}}{\pi}}\sqrt{\left(t_C - \frac{Z}{u_S}\right)}dz$$

where r is the inner radius of the limb, which is cylindrical in shape.

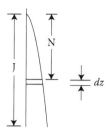

FIGURE 2.2 Self-draining falling film.

Mass transfer of oxygen in one limb of the bioreactor is found by integrating for the total height of oscillation. The expression takes the following form:

$$Gw = 4\sqrt{2}\pi r(Cai - Ca0)\sqrt{\frac{D_{ab}}{\pi}} \int_0^{u_S - t_c} \sqrt{\left(t_C - \frac{z}{u_S}\right)} dz$$

On integration, Gw takes the following form:

$$Gw = \frac{8\sqrt{2}}{3}\pi r(Cai - Ca0)\sqrt{\frac{D_{ab}}{\pi t_C}} u_S t_C^2 \tag{2.3}$$

Oxygen transfer to the flat surface The surface mass transfer through the flat surface of one limb of the bioreactor in one cycle of operation is given by the following form:

$$Gs = \pi r^2 (Cai - Ca0)2t_C \sqrt{\frac{4D_{ab}}{\pi 2 t_C}}$$

$$= 2\sqrt{2}\pi r^2 (Cai - Ca0)t_C \sqrt{\frac{D_{ab}}{\pi t_C}} \tag{2.4}$$

Total oxygen transfer in the bioreactor The total oxygen transfer to the equipment over a cycle of operation is calculated. The assumptions are

1. Both the cylinders are identical and operated under near-identical conditions over a cycle.

2. The radius of the vessels is large compared to the film thickness, and therefore the flat film analysis is valid.

3. The holdup of the system does not change significantly during operation.

4. The transfer rate is uniform.

5. At the completion of every cycle a uniform bulk liquid oxygen concentration is achieved in the bioreactor medium.

6. Even though bulk oxygen concentration varies from cycle to cycle, during the cycle the concentration is assumed to be constant.

7. The oxygen concentration in the gas phase is constant. Also, the saturation concentrations at the films and the flat liquid surface are the same and uniform during the operation.

8. Diffusivity of oxygen in the liquid is assumed to be unaffected by oxygen concentration and microbiological effects.

9. The process is isothermal.

The total oxygen deposited (in g/cycle) in the bioreactor in one cycle is given by

$$Gt = 2(Gw + Gs)$$
$$= \frac{16\sqrt{2}}{3}\pi r(Cai - Ca0)\sqrt{\frac{D_{ab}}{\pi t_C}}u_s t_C^2$$
$$+ 4\sqrt{2}\pi r^2(Cai - Ca0)t_C\sqrt{\frac{D_{ab}}{\pi t_C}}$$

The amount of oxygen deposited per unit volume per cycle in the bioreactor is given by Gt/v, where v is the working volume of the bioreactor. The average transfer rate (in g/s) is given by

$$Gt_a = \frac{Gt}{t_C}$$
$$= \frac{8\sqrt{2}}{3}\pi r(Cai - Ca0)\sqrt{\frac{D_{ab}}{\pi t_C}}u_s t_C$$
$$+ 2\sqrt{2}\pi r^2(Cai - Ca0)\sqrt{\frac{D_{ab}}{\pi t_C}}$$

And mass transfer coefficient in liquid phase K_L (in cm/s) is represented by the following expression:

$$K_L = \frac{Gt_a}{\left[(2\pi r^2 + 2\pi r u_s t_C)(Cai - Ca0)\right]}$$

$$= \frac{Gt_a}{2\pi r(r + u_s t_C)(Cai - Ca0)}$$

$$= \frac{1}{(r + u_s t_C)}\sqrt{\frac{D_{ab}}{\pi t_C}}\left(\frac{4\sqrt{2}}{3}u_s \cdot t_C + \sqrt{2}r\right) \qquad (2.5)$$

2.3 Experiments to verify the modeling

During experimentation u_s (i.e., the velocity of the falling/rising liquid column) is kept constant. Oxygen deposition in the bioreactor medium is recorded for different time periods of oscillation of the liquid column.

This is compared with the theoretical dissolved oxygen profile using the above formulation. Figures 2.3 through 2.5 represent the theoretical and experimental profiles of dissolved oxygen in the bioreactor for periods of oscillation of 20, 25, and 35 s, respectively.

The dotted lines represent the theoretical dissolved oxygen concentration, and the firm lines represent the actual recorded dissolved oxygen concentration with repetitive oscillations. The theoretically obtained and measured K_L values are given in Table 2.1.

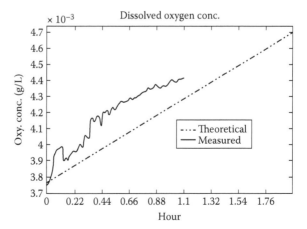

FIGURE 2.3   Theoretical and actual dissolved oxygen profiles for time period of oscillation = 20 s.

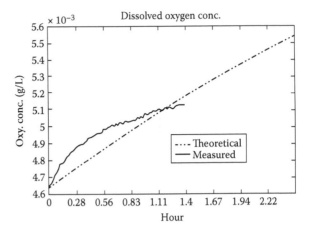

FIGURE 2.4 Theoretical and actual dissolved oxygen profiles for time period of oscillation = 25 s.

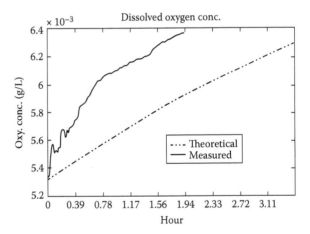

FIGURE 2.5 Theoretical and actual dissolved oxygen profiles for time period of oscillation = 35 s.

Table 2.1 Theoretical and actual K_L values

Exp. no.	K_L (theor.) (cm/s)	K_L (actual) (cm/s)	K_L (actual)/K_L (theor.)
1	0.00017409	0.00045683	2.6241
2	0.00015788	0.00033864	2.1449
3	0.00013592	0.00065032	4.7846

2.4 Discussion of the results

Figure 2.3 displays the actual and theoretical dissolved oxygen profiles in the bioreactor when the time period of oscillation is 20 s. This figure shows that the actual dissolved oxygen concentration in the bioreactor is higher than that of the theoretical one. The discrepancy may be explained as follows:

- In the theoretical expression, it was assumed that there is no turbulence in the flat surface of the reactor, but some turbulence was present in the liquid medium due to oscillation. This could be an important reason for the difference in the time profiles.

Figure 2.4 displays the actual and theoretical dissolved oxygen concentration profiles in the bioreactor when the time period of oscillation is 25 s. The figure displays an initially higher actual oxygen deposition, but at the end the actual profile shows drooping characteristics. The same may be explained as below:

- Apart from turbulence effects, the drooping nature of the actual profile can be explained by the presence of microorganisms in the bioreactor. The amount of oxygen deposited is being consumed by microorganisms present in the bioreactor medium.

Figure 2.5 shows that the actual oxygen deposition is much higher than the theoretical one. This is explained as below:

- When the time period of oscillation is 35 s, the liquid column rises to a great height and acquires great falling velocity, giving rise to turbulence. So, the amount of oxygen deposited is much higher than that theoretically predicted.

It is observed that the model of oxygen transfer in the equipment is fairly approximate. The several assumptions mentioned earlier may not be true. The system controller is developed later based on the actual dissolved oxygen measurement.

2.5 Future scope of work

The fabricated bioreactor under study is a prototype model. There exists a lot of work to be done in both design and fabrication aspects. Further studies can also be carried out for the theoretical model of gradation for more accurate prediction of the dissolved oxygen concentration.

1. The present theoretical model for oxygen deposition in the bioreactor medium could be improved by incorporating a turbulence effect.

2. Oxygen deposition in the bioreactor can be improved by installing extended surface configuration structures inside the cylinder.

3. Studies on a packed bed configuration can be done.

4. The oxygen deposition inside the bioreactor can be enhanced by a mixed-mode operation. That is, see-saw action can be accompanied by continuous bubbling through the bioreactor or a packed bed system as above.

5. Also, certain modifications in the fabrication of the present prototype bioreactor can streamline the working of the bioreactor.

It is observed from the experimentation that there remains a scope of improvement of the present model. However, the model developed withstood scrutiny. The fabricated bioreactor is sufficiently airtight to maintain the sterile condition for long fermentation work. Sterilization of the bioreactor can be done by passing steam through the steam jacket, with the steam in turn being derived from an elementary-scale boiler. Instrumentation for automated operation, including process and environmental control, worked satisfactorily in the present fabricated system. Yeast was successfully fermented in this bioreactor and thus examined the performance of the instruments and the controllers. The results are satisfactory.

It is concluded that this system could be used for animal cell line culture.

Further studies should be carried out regarding a higher mass transfer of oxygen in the bioreactor system.

CHAPTER THREE

Simulation of bioprocess and development of BIPROSIM

A general purpose simulation program

Biotechnology is the key for many types of products in pharmaceutical, food and beverage, and fermentation industries. To increase productivity and at the same time save raw material, energy, and time, the optimum operating mode of the bioreactor should be found. Although research efforts have been made in the direction of strain improvement of the bacteria or species of interest by the application of gene technology, efforts to adopt proper operational modes and optimal values of control inputs can hardly be ignored [23]. In this study we develop a systematic approach for selection of the best operating mode for a particular process for an interval during the fermentation. This is achieved by the following ones:

- Suitably modeling the bioprocess, an extremely complex task [24]

- Formulating different and possible operating modes for the model

- Developing a suitable algorithm that can predict the best combinations of operating modes for a bioreactor for a particular product, by simulation experiments

Bioengineers obtain information regarding the selection of a practical operating mode from biochemical experiments. These experiments are time-consuming and expensive. The algorithm described in this chapter saves time and expenditure in reaching a conclusion about the selection of appropriate operating modes.

The algorithm is user-friendly and completely menu driven such that the user can extract the necessary information without going into the formulation of the model. The package also displays the time profile of the bioprocess variables (BPVs) of interest. An application-specific session has been reported in Section 3.5.

3.1 Mathematical formulation of the bioprocess

Schematically, a bioreactor may be represented as in Figure 3.1.

We have made certain simplifying assumptions in modeling the bioprocess:

- The model used here has been developed using mass balance equations in liquid medium. This is an unstructured and unsegregated model.
- Composition inside the bioreactor has been considered totally homogeneous with substrate limitations.
- Monod's model has been used to represent growth kinetics.

The generalized unstructured, unsegregated model obtained from mass balance equations in liquid condition [1,25–28] is of the following form:

$$\frac{d\mathbf{m}(t)}{dt} = q_0(t)m_0(t) + q_4(t)m_4(t) - q_1(t)m_1(t) + g[\mathbf{m}(t)] \quad (3.1)$$

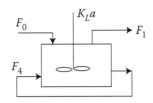

FIGURE 3.1 Schematic functional diagram of see-saw bioreactor.

where

g[**m**(t)] = material accumulation and consumption rate that describe biological activities

$q_0(t) = (F_0(t)/v(t))$ = normalized inflow to the bioreactor

$q_1(t) = (F_1(t)/v(t))$ = normalized outflow from the bioreactor

$q_4(t) = (F_4(t)/v(t))$ = normalized recycle flow to the bioreactor

m(t) = BPVs like cell mass, substrate, oxygen, and product concentrations

$v(t)$ = working volume of the bioreactor

More detailed expression of Equation 3.1 for some specific cases can be as below [13]:

$$\frac{dx_1(t)}{dt} = \frac{\mu_m x_1(t)x_2(t)x_3(t)}{(k_s + x_2(t))(k_c + x_3(t))} - k_d x_1(t) + \frac{F_0(t)}{x_5(t)}x_{in}$$

$$- \frac{F_1(t)}{x_5(t)}x_1(t) + \frac{F_4(t)}{x_5(t)}x_{11}$$

$$\frac{dx_2(t)}{dt} = -\frac{\mu_m x_1(t)x_2(t)x_3(t)}{Y(k_s + x_2(t))(k_c + x_3(t))} - m_S x_1(t) - \frac{\alpha x_1(t)}{Y_P}$$

$$- \frac{\beta\mu_m x_1(t)x_2(t)x_3(t)}{Y_P(k_s + x_2(t))(k_c + x_3(t))} + \frac{F_0(t)}{x_5(t)}S_{in} - \frac{F_1(t)}{x_5(t)}x_2(t)$$

$$+ \frac{F_4(t)}{x_5(t)}x_{12}$$

$$\frac{dx_3(t)}{dt} = -\frac{\mu_m x_1(t)x_2(t)x_3(t)}{Y_0(k_s + x_2(t))(k_c + x_3(t))} - m_{S0} x_1(t) - \frac{\alpha x_1(t)}{Y_{P0}}$$

$$- \frac{\beta\mu_m x_1(t)x_2(t)x_3(t)}{Y_{P0}(k_s + x_2(t))(k_c + x_3(t))} + \frac{F_0(t)}{x_5(t)}O_{in} - \frac{F_1(t)}{x_5(t)}x_3(t)$$

$$+ \frac{F_4(t)}{x_5(t)}x_{13} + K_L a(O_2^* - x_3(t))$$

$$\frac{dx_4(t)}{dt} = \alpha x_1(t) + \frac{\beta\mu_m x_1(t)x_2(t)x_3(t)}{(k_s + x_2(t))(k_c + x_3(t))} + \frac{F_0(t)}{x_5(t)}P_{in}$$

$$- \frac{F_1(t)}{x_5(t)}x_4(t) + \frac{F_4(t)}{x_5(t)}x_{14}$$

$$\frac{dx_5(t)}{dt} = F_0(t) - F_1(t)$$

where

$x_1(t)$, $x_2(t)$, $x_3(t)$, and $x_4(t)$ denote concentrations of cell mass, substrate, oxygen, and product, respectively, in the liquid phase of the bioreactor in g/m^3.

$x_5(t)$ denotes the working volume of the bioreactor in m³. The time unit is measured in hours.

Referring to Figure 3.1, F_0, x_{in}, s_{in}, o_{in}, and F_4 represent the liquid feed rate (feed-in rate), withdrawal rate (feed-out rate), and recycle rate in the bioreactor, respectively; x_{in}, S_{in}, O_{in}, and P_{in} are the influent cell mass, substrate, oxygen, and product concentrations, respectively; K_L denotes oxygen mass transfer coefficient; a is the surface area through which mass transfer of oxygen is taking place; k_S and k_{CO} represent the saturation constants; k_d, m_S, m_{S0}, Y, Y_0, Y_p, and Y_{p0} represent biomass decay rate, maintenance coefficients with respect to carbon and oxygen source, yield coefficients with respect to carbon and oxygen source for cell mass growth, and yield coefficients with respect to carbon and oxygen source for product formation, respectively; O_2^* represents saturation value of oxygen in liquid medium of interest; α, β are constants; $\mu'(x, t)$ represents the specific growth rate which can be assumed to have bounds as $0 \leq \mu \leq \mu_m$ for all x; μ_m is representing the maximal growth capacity.

3.2 Modes of operation

Different combinations of F_0, F_1, and F_4 define different operating modes for the model, as in Table 3.1. Note that $K_L a$ remains constant for all these operating modes.

In addition to the eight fixed operating modes in Table 3.1, the following two combination modes are also simulated.

Table 3.1 Different op-modes

F_0	F_1	F_4	Name op-modes	Remarks
0	0	0	Op-mode 1	Batch mode
0	0	1	Op-mode 2	—
0	1	0	Op-mode 3	—
0	1	1	Op-mode 4	—
1	0	0	Op-mode 5	Fed-batch mode
1	0	1	Op-mode 6	—
1	1	0	Op-mode 7	Continuous mode
1	1	1	Op-mode 8	—

Note: 1 stands for pump ON, which causes in-, out-, or recycle flow; 0 stands for pump OFF, blocking in-, out-, or recycle flow.

Op-mode 9   The bioprocess is simulated for all the operating modes (op-mode 1 through op-mode 8) for an arbitrary starting condition. The op-mode that produces the maximum BPV of interest or yield is selected for the first hour. The end result of first hour serves as the initial condition for the second hour and the procedure is repeated. Thus, a sequence of operating modes can be derived that will maximize the yield over the total fermentation period.

Op-mode 10   In this mode of operation we determine (of the nine operating modes) which will give the maximum and minimum BPVs of interest, respectively. Corresponding time profiles of the BPVs are displayed for the time span of fermentation.

3.3   Adoption in the model parameters

Parameters that may vary with time are yield coefficients Y, Y_0, Y_p, and Y_{p0} of the oxygen saturation value (O_2^*), maintenance coefficients m_S and m_{S0}, and the death rate coefficient (k_d). Since the parameters vary slowly, using suitable estimation algorithms (e.g., an extended Kalman filter) [29,30], it is possible to estimate them. The model parameters could be updated at regulated intervals.

3.4   The algorithm of BIPROSIM

As discussed earlier, the algorithm is menu driven. After asking for the selection of a bioreactor type (Figure 3.2), op-mode (Table 3.1), and product (Figure 3.3), the time profiles of various BPVs are displayed. This is done by taking default or standard

1. CSTR

2. Bubble column

3. Airlift

FIGURE 3.2   Types of bioreactors.

```
1. Yeast
2. Lactic acid
3. SCP
4. Erythromycin
5. Wine
6. Cephalosporin C
7. Alpha-amylase enzyme
```

FIGURE 3.3   Products considered here.

values of control inputs and parameters. Op-mode 9 asks for the BPV of interest, the one that is to be maximized. Op-mode 10 generally serves as the conclusion. Users can change the total fermentation time, starting BPVs, flow rates, $K_L a$ value, inflow, and recycle, and see their influence on the time profile displays. If this software is used for real-time application, the updated time-varying parameters are incorporated in the model. Thereafter, op-mode 9 serves to display the best operating mode for the current situation. A flowchart of the algorithm is shown in Figure 3.4. It provides necessary information about the best possible op-mode for the next hour. For Figure 3.4 x_{11}, x_{12}, x_{13}, and x_{14} are recycled cell mass substrate, oxygen, and product concentration respectively.

At the end, all the variables are reset to their standard or default values, as at the beginning. Thus, complete simulation of a bioprocess and detection of the best op-mode are possible using BIPROSIM, which is based on MATLAB®.

3.5 Sample run

A single cell protein (SCP) fermentation process has been chosen to do a sample run of BIPROSIM. The mathematical model for SCP fermentation can be described by the model equations discussed in Section 3.1, with the following starting parameter values:

$$\mu_m = 0.6102; \quad Y = 0.6; \quad Y_0 = 0.7; \quad Y_P = 0.62;$$
$$Y_{p0} = 0.6; \quad k_s = 0.31; \quad k_c = 0.01$$
$$m_s = 0.01; \quad m_{s0} = 0.00001; \quad k_d = 0.001; \quad \alpha = 0.01; \quad \beta = 0$$

The following values are also selected.

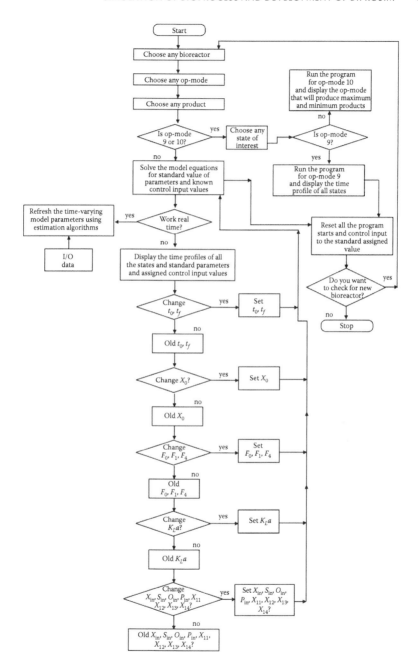

FIGURE 3.4   Flowchart of BIPROSIM.

Case 1:

Total time of fermentation $= 25$ h

$$K_L a = 0.0128; \quad F_0 = 0.003; \quad F_1 = 0.005; \quad F_4 = 0.002;$$

Concentrations of different BPVs along with inflow, outflow, and recycle flow are chosen as

$$x_{in} = 0; \quad S_{in} = 10; \quad O_{in} = 0.008; \quad p_{in} = 0; \quad x_{11} = 0.1;$$
$$x_{12} = 0.5; \quad x_{13} = 0.008; \quad x_{14} = 0.1$$

Assuming the above-mentioned standard values, the time profiles of all the BPVs in continuous stirred tank reactor (CSTR), assuming op-mode 6 for SCP fermentation, is shown in Figure 3.5.

In the second case the values of the variables have been changed as under

Case 2:

Total time of fermentation $= 30$ h

$$f_0 = 0.06; \quad f_4 = 0.12; \quad K_L a = 0.005$$

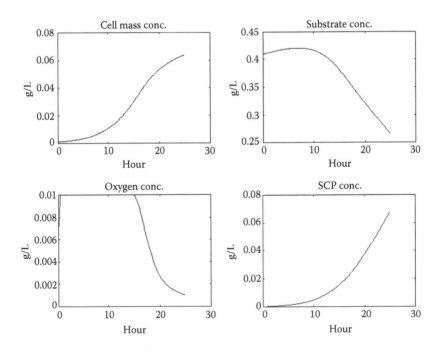

FIGURE 3.5   SCP fermentation under given conditions in op-mode 6 using BIPROSIM.

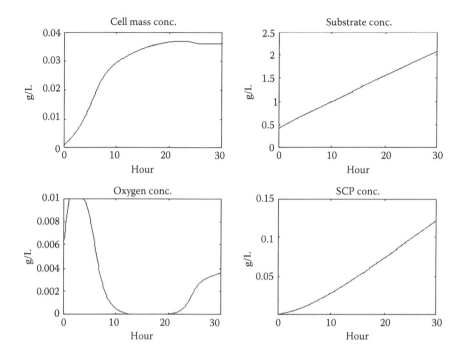

FIGURE 3.6   SCP fermentation under chosen conditions in op-mode 6 using BIPROSIM.

The resultant time profiles of all the BPVs are displayed in Figure 3.6.

If we run op-mode 9 and want to maximize BPV $x(4)$ (product concentration), the following sequence of operations has been calculated.

The bioreactor should run for full 30 h in op-mode 6. The resultant time profiles of all the BPVs are shown in Figure 3.7.

The results obtained after running op-mode 10 are as follows:

Operating mode that will produce the maximum product ($x(4)$) is op-mode 9.

Concentration of product ($x(4)$) is calculated to be 1.0334 g/L.

Operating mode that will produce the minimum concentration of ($x(4)$) is op-mode 1.

Concentration of product ($x(4)$) is calculated as 0.0636 g/L.

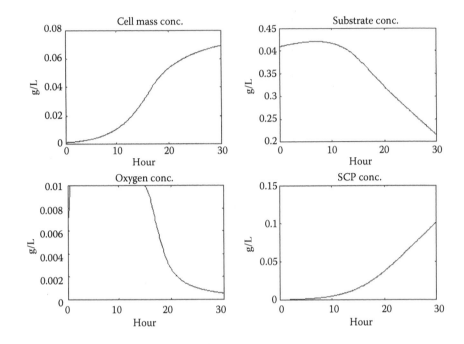

FIGURE 3.7   SCP fermentation under given conditions in op-mode 9 using BIPROSIM.

Case 3:
We run the program with the following changed conditions:
Total time of fermentation = 36 h

$$f_0 = 0.03; \quad f_1 = 0.09; \quad f_4 = 0.08; \quad K_L a = 0.02$$

The BPV of interest is product concentration ($x(4)$).

Op-mode 9 indicates that the following sequence of operation is to be executed to maximize product concentration ($x(4)$).

First hour, the bioreactor should run in op-mode 2.

It should run in op-mode 6 for the next 10 h.

Then it should run in op-mode 2 for the last 25 h.

The time profiles of the BPVs are shown in Figure 3.8.

Op-mode 10, in this situation, will produce the following results:

Operating mode that will produce the maximum product is op-mode 9.

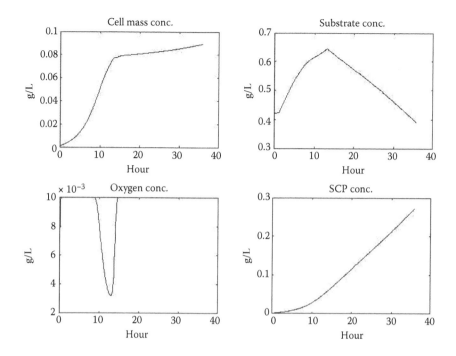

FIGURE 3.8   SCP fermentation under chosen conditions in op-mode 9 using BIPROSIM.

Concentration of product ($x(4)$) is calculated to be 1.0224 g/L.

Operating mode that will produce the minimum product is op-mode 1.

Concentration of product ($x(4)$) is calculated to be 0.1356 g/L.

Similar simulations can be carried out for other products. A software tool for finding the optimum operating mode or a sequence of operating modes of a bioreactor, to maximize the production of a BPV of interest, has been reported. The best and worst operating modes for a particular process can also be predicted. For example, op-mode 6 (fed batch) should not be used for lactic acid fermentation.

Thus, the software package BIPROSIM will score as a useful tool for maximizing production in fermentation industries. This will aid a researcher in making the optimum choice of op-modes in operating a bioreactor for a particular product. The time span up to which fermentation will be profitable can also be decided by this package. Later on we calculate optimal control input time profiles using this package, with some modifications for generating the optimal time profiles of the BPVs. This will further improve the production rate. This is discussed in Chapter 4.

CHAPTER FOUR

Dynamic optimization of a bioprocess using genetic algorithm

This chapter is concerned with maximizing the product concentration in a fermentation process. This is achieved by the selection of flow rates to the bioreactor and a sequence of mixed modes of operations. Since bioprocesses never reach a steady state in their operating time span, the optimization technique is a dynamic one. The bioprocess model used is nonlinear with constraints. A genetic algorithm (GA) has been used for dynamic optimization of the process. The results serve as optimal time profiles of process variables and act as reference time profiles for the controller.

4.1 Historical background

Maximizing the product concentration in a fermentation process has been the subject of research for quite some time. The objective of this chapter is to maximize the product concentration by a combination of simulation experiments and an optimization procedure. An increase in product concentration leads to larger productivity and profitability for the fermentation industry.

Of the different methodologies for achieving the same goal, one is strain improvement using gene technology. An alternative is optimized feed scheduling.

With present-day computational facilities, simulation experiments provide a good substitute for laboratory experimentation [25–27]. The outcome can be verified later by actual experimentation. In this chapter, a two-pronged strategy has been adapted for higher product yield. A combination of different operating modes [31], along with optimized flow rates, $K_L a$, and so forth, has been used in the simulation experiment. The optimization is based on GAs [32] with suitable alterations for adaptation to bioprocess optimization. It has been found that considerable improvement is possible by the above two combinations.

Development of the bioprocess model is a prerequisite for simulation. In Chapter 3, formulation of the bioprocess model has been presented and the different possible operating modes of the bioreactor have been listed.

GA has been applied to find the optimum control inputs under different operating modes. Results showing the time profiles of different bioprocess variables and the corresponding control variables obtained from the simulation experiment are presented.

4.2 Bioprocess model development

The generalized model for a bioreactor system has been discussed in Chapter 3. It is an unstructured and unsegregated model, and is the outcome of mass balance in the liquid medium of the reactor [2,33]. Different possible operating modes have been given in Table 3.1.

Structured and segregated models [34] have not been used in the present case, as they are far more complicated with a large number of parameters and states, many of which are unknown and unmeasurable. They are not suitable for optimization.

The model equations of Chapter 3 can be written in matrix form as

$$\dot{\mathbf{x}}(t) = \mathbf{f}(\mathbf{x}, t) + H\mathbf{x}(t) + B(\mathbf{x}, t)\mathbf{u}(t) \tag{4.1}$$

$$\dot{x}_5(t) = F_0(t) - F_1(t) \tag{4.2}$$

where $\mathbf{x}(t)$ denotes the state vector,

$$\mathbf{x}(t) = [x_1(t) x_2(t) x_3(t) x_4(t)]^T$$

and $\mathbf{u}(t)$ denotes the input vector,

$$\mathbf{u}(t) = [F_0(t)F_1(t)F_4(t)(K_L a(t))]^T$$

The constraints are as follows:

$$0 \le x_1(t) \le 60 \quad 0 \le F_0(t) \le 1.6$$

$$0 \le x_2(t) \le 100 \quad 0 \le F_1(t) \le 1.6$$

$$0 \le x_3(t) \le 0.01 \quad 0 \le F_4(t) \le 1.6$$

$$0 \le x_4(t) \le 0.15 \quad 0 \le K_L a(t) \le 0.0168$$

Other parameters, like S_{in}, O_{in}, x_{11}, x_{12}, x_{13}, and x_{14}, are also bounded as below:

$$0 \le S_{in}(t) \le 100$$

$$0 \le O_{in}(t) \le 0.01$$

$$0 \le x_{11}(t) \le 0.006$$

$$0 \le x_{12}(t) \le 30$$

$$0 \le x_{13}(t) \le 0.003$$

$$0 \le x_{14}(t) \le 0.0001$$

The aim of this chapter is to find the optimum time profiles of F_0, F_1, F_4, and $K_L a$ under a particular operating mode such that the yield $x_4(t)$ is maximized. Finally, optimized time profiles of cell mass concentration (x_1), substrate concentration (x_2), dissolved oxygen concentration (x_3), and product concentration (x_4) are obtained. These will serve as optimized time profiles and will be used as reference profiles for the purpose of control.

4.3 Application of genetic algorithm for control input optimization

The nonlinear bioprocess model, along with the imposed state and control constraints, makes the solution of the dynamic optimization problem a difficult one by existing classical techniques. GAs have been used for solving the problem of product yield maximization. These have many advantages compared to classical optimization procedures. First, GA is robust. It works from a rich database of points simultaneously (i.e., from a population of strings) climbing many peaks in parallel; thus the possibility of getting stuck at a false peak is reduced over methods that go point to point. In short, GA ensures that the result obtained will be the global maximum.

The bioprocess usually operates for hours. The GA is made to operate on a sampling instant-to-instant basis for operating modes 1 through 8. At the end of each sampling instant, the optimal control inputs as well as optimal state values are stored. For the next instant calculations, initial values of the states are updated. The control matrix is updated by making use of the optimum control input of the previous results. This is repeated for the whole operating period. For mode 9, the above-mentioned process is repeated for an hour.

Algorithm 4.1, along with subroutines "Child" and "Mutate," indicates how GA has been applied to find optimum control inputs instant by instant.

Algorithm 4.2 indicates how GA has been used to find optimum control inputs hour by hour for the whole of the working time span. Flowcharts 4.1 and 4.2 correspond to Algorithms 4.1 and 4.2, respectively.

For the present example, the following are the assumptions:

1. Total time of operation is 18 h.
2. Sampling interval is 6 min.
3. The only measurable parameter is the dissolved oxygen concentration.

The fitness function of different op-modes has been assigned as below from the viewpoint of engineering judgment.

- Fitness function of op-mode 1 = $2000x_4(t)$
- Fitness function of op-mode 2 = $2000x_4(t)$
- Fitness function of op-mode 3 = $500x_4(t) + (x_5(t) - 3)$
- Fitness function of op-mode 4 = $500x_4(t) + (x_5(t) - 3)$

- Fitness function of op-mode 5 = $2000x_4(t) + (15 - x_5(t))$
- Fitness function of op-mode 6 = $2000x_4(t) + (15 - x_5(t))$
- Fitness function of op-mode 7 = $2000x_4(t) + (1/(13 - x_5(t)))$
- Fitness function of op-mode 8 = $2000x_4(t) + (1/(13 - x_5(t)))$

Recall, x_5 is the reactor working volume in m³. If the oxygen concentration exceeds the prescribed limit, it is thresholded to 0.01 g/L at each sampling instant.

4.4 Explanation of the application of GA

Recall that Algorithm 4.1 shows how to calculate the optimum control inputs for one sampling instant. Algorithm 4.2 gives the implementation of GA for the whole time span and displays the optimum time profile of the process variables and inputs.

Algorithm 4.1 A seed is selected arbitrarily. Ten sets of random numbers are generated within the physical bounds of control inputs. A (10 × 10) control input matrix is formed.

For any operating mode, values of the fitness function are determined corresponding to the 10 control inputs of the control matrix. This is done by solving the model equations discussed in Section 3.1 corresponding to each control input set. The control input that produces the maximum fitness value is noted.

In the next generation, a control matrix is formed in the following way. The control input set that generates the highest fitness value is placed in the first row; the remaining nine rows are filled by offspring generated using the reproduction technique of the GA. This is done by using the subroutine "Child." Any two rows are selected randomly. They are copied in a file. A random number is picked up between 0 and 4 (say r_1). Crossover is forced across the rth column of two rows. Thus two new offspring are produced. One of them is selected at random and placed in the control matrix. Thus all the nine rows are replaced.

Sometimes, "Mutation" is used instead. This is done by the subroutine "Mutate." Here, a partial control matrix is formed of dimension ($r_2 \times 10$), where r_2 is a random number between 1 and 9. Another random number r_3 is generated whose value is either 0 or 1. If $r_3 = 0$, the second to ($r_2 + 1$)th row of the control matrix is replaced by the partial control matrix. Otherwise, the control matrix is retained.

If process variables like dissolved oxygen concentration and volume exceed the specified limit, the fitness value of the corresponding control input is set to a very low value (minus infinity). This helps in satisfying the state constraint. If the state constraint is still not met, the corresponding control input set is replaced. This is done until all the state constraints are met.

At the end of each generation, the maximum fitness value is stored. At the end of the fifth generation, the average of the five fitness values is calculated. The difference between the average fitness value and the fitness value of the last generation (which represents the best value) is calculated. This is the error. If the computed error exceeds some predefined set value, the process is continued. Alternatively, the best control input of the last generation represents the optimum control input.

Flowchart of Algorithm 4.1

- Step 1: Select *seed* = 1619295.
  i. Generate random numbers within the upper bound of the control inputs.

$$(F_0, F_1, F_4, K_L a, S_{in}, O_{in}, x_{11}, x_{12}, x_{13}, x_{14})$$

  ii. Generate 10 such sets of numbers to form the control matrix.
- Step 2: Define fitness function.
- Step 3: Set *error* = 1.
- Step 4: If *error* < 0.0000001, go to step 19.

Otherwise go to step 5.

- Step 5: Set *generation* = 1.
- Step 6: Define the (10×10) matrix generated in step 1 or else as the control matrix.
- Step 7: Find the fitness value corresponding to each row of the control matrix. Discard the control input set for which the fitness value reaches infinity.
- Step 8: If {(dissolved oxygen concentration and volume) > specified limits}

Set *fitness value* = −*infinity*
else, go to step 9.

- Step 9: While {(dissolved oxygen concentration and volume) > specified limits}, change the control inputs with new random numbers until constraints are satisfied.

- Step 10: For the formation of the second-generation control matrix,
    - i.   The control inputs that generate the maximum fitness value are placed in the first row.
    - ii. To fill the remaining rows, call the subroutine "Child."
- Step 11: To implement mutation in the control matrix, call the subroutine "Mutate."
- Step 12: Control matrix = new control matrix.
- Step 13: Generation = generation + 1.

Replace the value of the generation in step 5 by this value.

- Step 14: If (generation $\leq$ 5), go to step 5.

Else go to step 15.

- Step 15: Find the fitness value for the fifth generation.
- Step16: Set *average value* = 99999.
- Step17: Define

error = average value – best fitness value
Replace the magnitude of error in step 3 by this value.

- Step 18: Calculate the average of the fitness values of five generations.

Replace the magnitude of the average value in step 16 by this value.

- Step 19: Print the result, that is, the optimum control inputs and the resultant process variables.
- Step 20: Stop.

The flowchart of Algorithm 4.1 is given in Figure 4.1.

Subroutine "Child"
- Step 1: Set i = 1.
- Step 2: Select any two rows of the control matrix randomly.
- Step 3: Copy them to a new file.
- Step 4: Generate a random number $0 < r_1 < 4$
- Step 5: Crossover is carried out around the r_1th column.
- Step 6: Two new offspring are produced.
- Step 7: Select any one of them randomly.

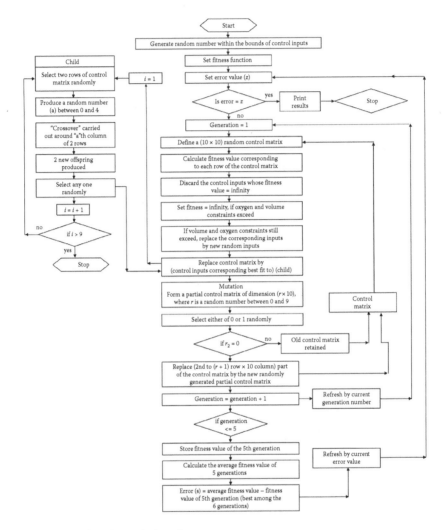

FIGURE 4.1   Flowchart of Algorithm 4.1.

- Step 8: $i = i + 1$.
- Step 9: If ($i > 9$), go to 10.
- Step 10: Stop.

Subroutine "Mutate"

- Step 1: Control matrix.
- Step 2: Generate a random number $1 < r_2 < 9$.
- Step 3: Generate a partial random matrix of dimension $(r_2 \times 10)$.
- Step 4: Generate a random number between 0 and 1.

- Step 5: If random number = 0, go to step 6; otherwise, go to step 1.
- Step 6: Replace the (2: $(r_2 + 1) \times 10$) part of the control matrix of step 1 by partial random control matrix.
- Step 7: $i = i + 1$.
- Step 8: If ($i > 9$), go to step 9; otherwise, go to step 2.
- Step 9: Stop.

Algorithm 4.2

The determination of the optimum control input profile for the whole working span is done here.

Starting from given inputs and a random control matrix, the optimum control input set for the first sampling instant is determined using Algorithm 4.1. For the next sampling instant, the end result of the first instant acts as the initial value. The control matrix is prepared in the following way. The first row of the control matrix is filled up by the best control input set of the previous instant. The rest of the control matrix is filled by random numbers (within specified bounds) and the process is repeated.

The best control input set of each sampling instant is stored. A profile of the optimum control input is obtained accumulating all the instantaneous results.

Flowchart of Algorithm 4.2

- Step 1: Set $j = 1$ (i.e., first sampling instant).
- Step 2: Set initial condition of states as

$$X_0 = [x_{10} x_{20} x_{30} x_{40}]$$

Control inputs from the random control matrix.

- Step 3: Apply Algorithm 4.1.

Find optimal control inputs for that sampling instant and store it. Also find the value of the process variables and store it.

- Step 4: Considering X_0 of step 2 and the optimum control inputs of step 3, solve the model equations.

Store the last result of the solution. After thresholding it, that is, if $x_3(t) < 0.00001$, put $x_3(t) = 0.00001$.

- Step 5: Replace X_0 of step 2 by the stored results of step 4.
- Step 6: Construct the control matrix as = [control inputs corresponding to the highest fitness value in the last iteration; [random numbers]].

Replace the control matrix of step 2 by the newly formed control matrix.

- Step 7: $j = j + 1$; replace the value of j in step 1.
- Step 8: If $j > 182$, go to step 9.

Otherwise, go to step 1.

- Step 9: Print the value of the optimized control inputs for all sampling instants along with that of the process variables.
- Step 10: Stop.

The flowchart of Algorithm 4.2 is given in Figure 4.2. Description of the figures displaying results of GA is shown in Table 4.1.

4.5   Results of the application of GA

GA has been applied to all the possible operating modes described in Table 3.1. Figures 4.3 through 4.20 illustrate the optimal time profiles of the states and control inputs, respectively, for the different operating modes.

For example, Figure 4.3 shows the optimal time profile of all the process variable in op-mode 9. Figures 4.4 through 4.6 show the optimized control inputs for op-mode 9. Similarly, Figures 4.7, 4.11, 4.13, and 4.16 through 4.20 depict the optimal time profiles of process variables of the same process for op-modes 8 through 1 as we count down in order of decreasing complexity of operation. Figures 4.8 through 4.10 depict the optimal control inputs for op-mode 8. Figure 4.12 depicts the optimal control inputs for op-mode 7, and Figures 4.14 and 4.15 depict the same for op-mode 6.

It has been found that the following sequence of operating modes will produce the maximum single cell protein (SCP) concentration for a fermentation time of 18 h (the values of the other parameters and initial conditions are kept the same as in Section 3.5): for the 1st, 2nd, and 3rd hours, op-mode 2; 4th hour, op-mode 6; 5th and 6th hours, op-mode 2; 7th hour, op-mode 4; 8th, 9th, and 10th hours, op-mode 6; 11th hour, op-mode 4; 12th hour, op-mode 2; 13th hour, op-mode 6; 14th hour, op-mode 2; 15th and 16th hours, op-mode 6; 17th hour, op-mode 2; 18th hour, op-mode 1.

The results show the applicability of GAs in solving the yield optimization problem in bioprocess models. By making

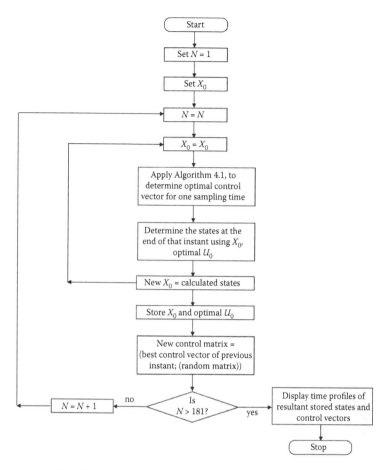

FIGURE 4.2 Flowchart of Algorithm 4.2.

Table 4.1 Description of the figures displaying results of GA

Op-mode	Figure displaying bioprocess variables	Figure displaying inputs
9	4.3	4.4, 4.5, 4.6
8	4.7	4.8, 4.9, 4.10
7	4.11	4.12
6	4.13	4.14, 4.15
5	4.16	—
4	4.17	—
3	4.18	—
2	4.19	—
1	4.20	—

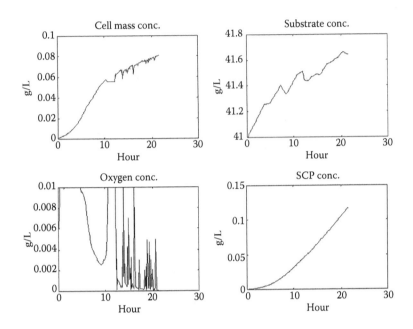

FIGURE 4.3   Optimum time profiles of SCP fermentation process under op-mode 9.

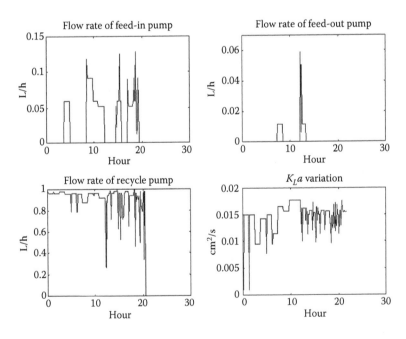

FIGURE 4.4   Optimum control input profiles of SCP fermentation process under op-mode 9.

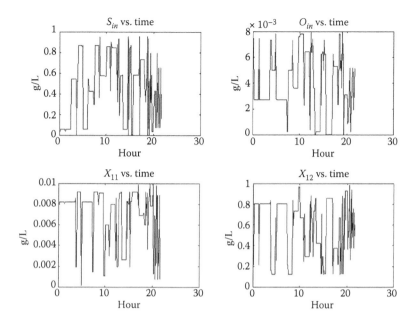

FIGURE 4.5 Optimum control input profiles of SCP fermentation process under op-mode 9.

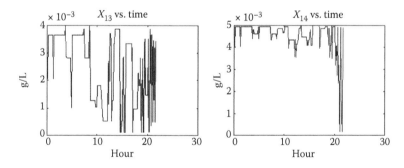

FIGURE 4.6 Optimum control input profiles of SCP fermentation process under op-mode 9.

the set value of error smaller and smaller, it is possible to obtain near-optimal results, and the results become independent of the seed. It will be very difficult to solve the problem by known classical techniques. As the result becomes independent of seed, the result obtained tends to be the global maximum.

In Chapter 3, how to find the optimal operating mode that will produce maximum yield was investigated. In this chapter, it is shown that GA can maneuver the flow and oxygen

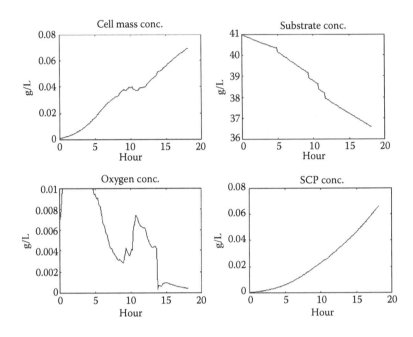

FIGURE 4.7   Optimum time profiles of SCP fermentation process under op-mode 8.

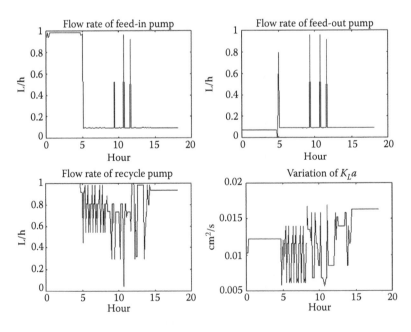

FIGURE 4.8   Optimum control input profiles of SCP fermentation process under op-mode 8.

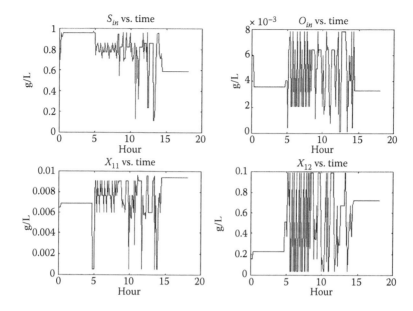

FIGURE 4.9 Optimum control input profiles of SCP fermentation process under op-mode 8.

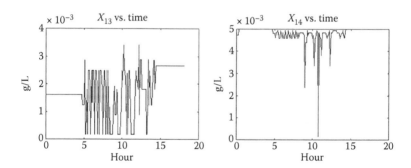

FIGURE 4.10 Optimum control input profiles of SCP fermentation process under op-mode 8.

concentration so as to produce the maximum output of products in the particular operating mode. So, we can decide on an optimal op-mode and maneuver the control inputs in that op-mode to maximize yield. The result obtained will ensure global maximization of the process variable of interest, and time profiles of other states will be optimal. BIPROSIM has GA incorporated in it. A flowchart of the modified BIPROSIM is shown in Figure 4.21.

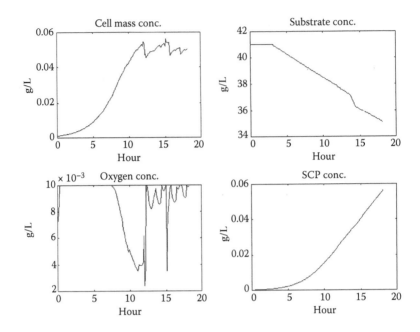

FIGURE 4.11   Optimum time profiles of SCP fermentation process under op-mode 7.

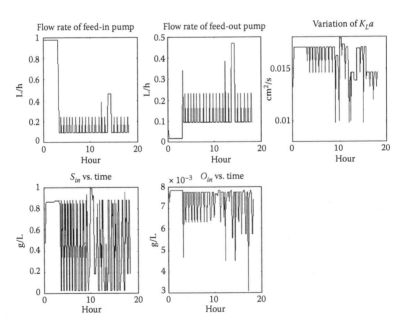

FIGURE 4.12   Optimum control inputs for SCP fermentation process under op-mode 7.

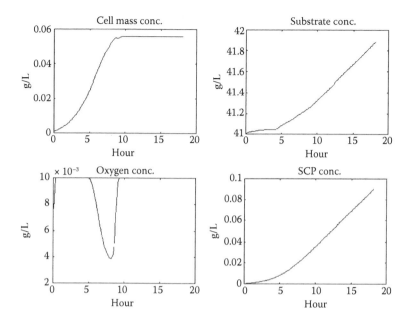

FIGURE 4.13 Optimum time profiles of SCP fermentation process under op-mode 6.

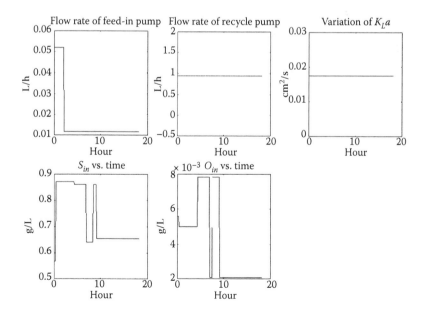

FIGURE 4.14 Optimum control inputs for SCP fermentation process under op-mode 6.

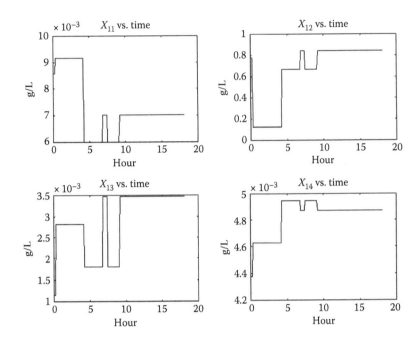

FIGURE 4.15   Optimum control inputs for SCP fermentation process under op-mode 6.

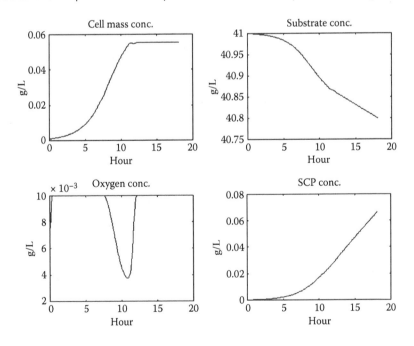

FIGURE 4.16   Optimum time profiles of SCP fermentation process under op-mode 5.

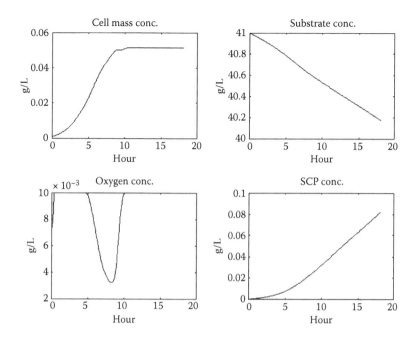

FIGURE 4.17 Optimum time profiles of SCP fermentation process under op-mode 4.

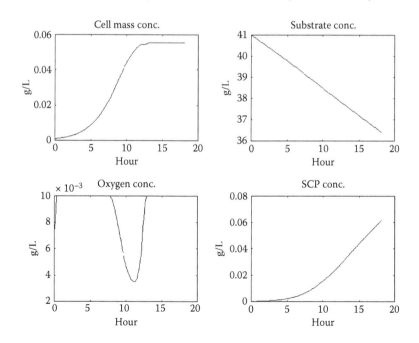

FIGURE 4.18 Optimum time profiles of SCP fermentation process under op-mode 3.

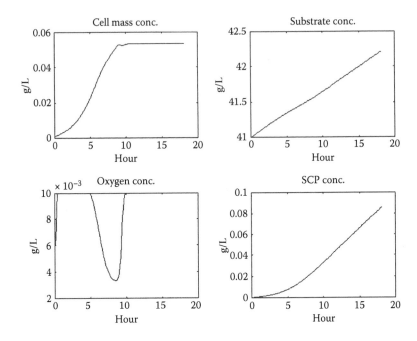

FIGURE 4.19   Optimum time profiles of SCP fermentation process under op-mode 2.

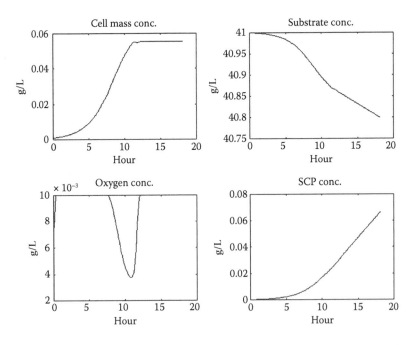

FIGURE 4.20   Optimum time profiles of SCP fermentation process under op-mode 1.

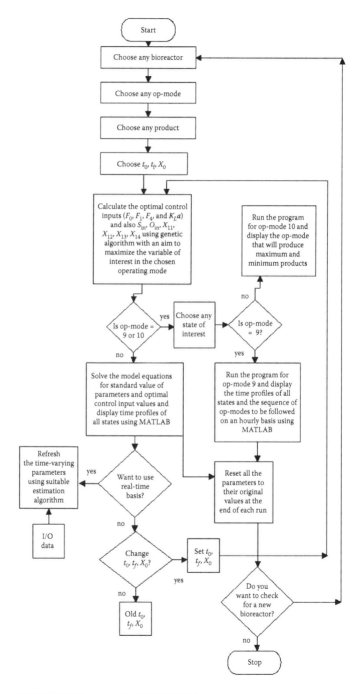

FIGURE 4.21 Modified BIPROSIM.

CHAPTER FIVE

Bioprocesses and time delay control

A bioprocess, as discussed earlier, is a time-varying, nonlinear, undermodeled multivariable system. Hence designing a bioprocess controller is not an easy task. Until now, proportional, integral, and derivative (PID) or adaptive-PID controllers have been used for bioprocess control purposes. However, these controllers, though working satisfactorily in the laboratory-scale model, do not work so in the scaled-up version. Instead, a time delay controller is expected to work satisfactorily for bioprocess control because it can take care of the effects of uncertainty and the undermodeling during the control operation of the process. Further, a time delay controller is computationally less complex than other sophisticated controllers. Hence scope of application of this type of controller for the control of bioprocesses is promising. In this chapter, a time delay controller and its application to control a bioprocess have been investigated. It is observed that the time delay controller is robust and its performance is comparable, if not superior, to the conventional PID controller.

5.1 The problem of bioprocess control

Biotechnological processes or, in short, bioprocesses, involve biochemical enzymatic reactions, organic and inorganic reactions, and increase in cell mass of the microorganisms in a suitable environment. These processes are marked by the following peculiarities:

1. The dynamics of these processes are poorly understood. The resultant model that is developed from the mass balance concept describes the process incompletely and is highly nonlinear.

2. The model parameters often vary with time due to metabolic variations and physiological and genetic modifications. These pose great difficulty in the control of these processes.

3. The reproducibility of the biotechnological experiments is uncertain.

Conventionally, bioreaction processes are classified as follows according to the mode of operation:

Batch: In the batch mode of operation neither the substrate is added to the initial charge, nor the product is removed until the end of the process. Some pharmaceutical preparations are made in this way.

Continuous: In the continuous mode of operation, substrate is continually added and product removed from the bioreactor. This is more economic than in the batch mode, for example, continuous fermentation of milk in the production of milk-based food and biological purification of wastewater.

Fed batch: The feed rate may be changed during the process, but no product is removed. This is the most economical mode of operation. Baker's yeast and antibiotics such as penicillin are made in the fed-batch mode commercially.

Recalling from Chapter 3 that there are nine operating modes of the bioprocess mentioned in Table 3.1, and they actually fall in between the above three conventional modes of operation.

The bioreactor system had been the subject of application of various linear, nonlinear, and adaptive control strategies by previous workers. Linearized control schemes generally fail or show poor performance. Even nonlinear controllers [35] do not work satisfactorily in the face of external uncertainties or unforeseen changes in the process. Various adaptive control strategies, linear or nonlinear, have been applied to impart robustness to the system performance in different operating modes against parametric variations with limited good results [36–40]. Sometimes good results of the benchtop model are not reproducible in scaled-up conditions.

The objective of this chapter is to present a new control methodology, namely, time delay control (TDC), for bioprocesses [41,42]. The unmeasurable process variables are estimated by a suitable observer [43,44].

The bioreactor is represented in a generalized way, having four control inputs: inflow, outflow, recycle flow, and oxygen transfer rate coefficient. The control objective is to achieve the maximum product yield. We take a three-pronged approach:

1. The optimum time trajectories of all the process variables are determined. This is described in Chapter 4 [16].

2. A time delay controller is used as a tracking controller such that the process variables are forced to follow the optimal time trajectories within finite error bounds.

3. As mentioned earlier, bioprocess variables are rarely measurable. These unmeasurable process variables are estimated by a suitable observer.

In addition to the environmental variables, temperature and pH of the bioreactor medium are closely regulated to a pre-defined value using suitable on–off or proportional controllers.

The difference between TDC and other controllers is that time delay controller has provision for incorporating unmodeled parts of the system dynamics and unforeseen disturbances, in addition to controlling nonlinearity. Also, TDC appears to be a much better control strategy from the viewpoint of computation. TDC is simpler, as it does not have to estimate parameters, as in the case of an adaptive controller. This makes it a candidate for real-time control.

Learning controllers such as fuzzy and neurofuzzy controllers have been tried in the case of bioprocess control [45]. This type of approach generally works well in repetitive processes, but in bioprocesses, repetition does not always produce the same results, thus making application of learning controllers rather ineffective.

To apply TDC, it is essential that all the states (and their derivatives) be available. Since most of the states are unmeasurable, a suitable observer is required.

This chapter is organized as follows. First, the required model of bioprocess for designing the time delay controller is presented in the state–space form. This formulation is as described in Chapter 3. The details of the time delay controller design are presented subsequently. In Section 5.4, application of TDC is described. In Section 5.5, the detailed development of a full-order observer compatible with TDC is described. The results and discussion of the simulation experimentation are presented in Section 5.7. The performance of conventional proportional integral (PI) and PID controllers has been compared to that of time delay controller.

5.2 Development of dynamic model

Three different types of models could be used for control purposes. The first is the physiological model, where knowledge of the physiology of the growth process is expressed in terms of antecedents and consequences. Strictly speaking, they are not mathematical models. Fuzzy control theory uses such physiological models, where a process operator's experience can be exploited in the controller design.

The second is the structured model [33,46,47], where a system of partial differential equations is used to characterize various internal states of the microorganisms. The variables are classified or structured according to the age of the cell, species of the organism, or various organic carbon compounds in the growth medium, secondary products, and so forth. The structured fermentation models are mathematically very complex. It is almost impossible to design a controller based on this model, as adequate sensing and controlling means are not available.

An unstructured and unsegregated model of bioprocesses takes into account that the fermentation is assumed to be dominated by a single, homogeneously growing organism [2,28]. In this chapter the third approach, that is, the unstructured model, has been considered for designing the controller. Monod's model with double-substrate limitations has been adopted to represent the growth kinetics [2] of different microorganisms of interest.

When the process takes place in a bioreactor with carbon substrate and oxygen supply, the dynamical model expresses the mass balance of the various components in the reactor, as discussed in Section 3.1.

Recalling $\mathbf{x}(t)$, denote the state vector:

$$\mathbf{x}(t) = [x_1(t) \; x_2(t) \; x_3(t) \; x_4(t)]^T$$

And recalling $\mathbf{u}(t)$, denote the input vector:

$$\mathbf{u}(t) = [F_0(t) F_1(t) F_4(t)(k_l a(t))]^T$$

$$\mu = \frac{\mu_m x_1(t) x_2(t) x_3(t)}{(k_d + x_2(t))(k_c + x_3(t))}$$

Represents Monod's model of specific growth rate and is evidently nonlinear and time-varying in nature. The above modeling is often uncertain. In spite of this, Monod's model represents a basic quantitative knowledge of μ. It is a well-known fact in bioengineering that the growth capacity of a population of microorganisms is intrinsically limited irrespective of environmental conditions. This means that the specific growth rate $\mu(x, t)$ can be assumed to have bounds as $0 \leq \mu \leq \mu_m$ for all x, with μ_m representing the maximal growth capacity.

It is assumed that dissolved oxygen is the only measurable quantity and constitutes the single output variable. Other bioprocess variables need to be estimated.

Additionally, we assume that for a constant input the model represented by the above-mentioned equations has a single asymptotically stable equilibrium point. This can be guaranteed by the reasonable restriction that the equilibrium point lies in the positive quadrant, that is, $x_1(t) > 0$, $x_2(t) > 0$, $x_3(t) > 0$, $x_4(t) > 0$, and $x_5(t) > 0$. It also implies that for constant input, stable states and the system matrix inverse define that the steady state exists. Another assumption is made that the process is bounded-input bounded-output (BIBO) stable.

The bioreactor model equations can be written in the following matrix form:

$$
\begin{bmatrix} \dot{x}_1(t) \\ \dot{x}_2(t) \\ \dot{x}_3(t) \\ \dot{x}_4(t) \end{bmatrix} = \begin{bmatrix} \mu \\ -\left(\dfrac{\mu}{Y} + \dfrac{\beta\mu}{Y_P}\right) \\ -\left(\dfrac{\mu}{Y_0} + \dfrac{\beta\mu}{Y_{P0}}\right) \\ \beta\mu \end{bmatrix} x_1(t) x_2(t) x_3(t)
$$

$$
+ \begin{bmatrix} -k_d & 0 & 0 & 0 \\ -\left(m_S + \dfrac{\alpha}{Y_P}\right) & 0 & 0 & 0 \\ -\left(m_{S0} + \dfrac{\alpha}{Y_{P0}}\right) & 0 & 0 & 0 \\ \alpha & 0 & 0 & 0 \end{bmatrix} \begin{bmatrix} x_1(t) \\ x_2(t) \\ x_3(t) \\ x_4(t) \end{bmatrix}
$$

$$
+ \begin{bmatrix} \dfrac{x_{in}}{x_5(t)} & -\dfrac{x_1(t)}{x_5(t)} & \dfrac{x_{11}}{x_5(t)} & 0 \\ \dfrac{S_{in}}{x_5(t)} & -\dfrac{x_2(t)}{x_5(t)} & \dfrac{x_{12}}{x_5(t)} & 0 \\ \dfrac{O_{in}}{x_5(t)} & -\dfrac{x_3(t)}{x_5(t)} & \dfrac{x_{13}}{x_5(t)} & (O_{sat} - x_3(t)) \\ \dfrac{P_{in}}{x_5(t)} & -\dfrac{x_4(t)}{x_5(t)} & \dfrac{x_{14}}{x_5(t)} & 0 \end{bmatrix} \begin{bmatrix} F_0(t) \\ F_1(t) \\ F_4(t) \\ (k_l a) \end{bmatrix} \qquad (5.1)
$$

$$\dot{x}_5(t) = F_0(t) - F_1(t) \tag{5.2}$$

$$y(t) = \mathbf{C}^T x(t) \tag{5.3}$$

where

$$\mathbf{C}^T = [0010]$$

Thus a multivariable nonlinear system with four numbers of control inputs and one output is defined by means of a finite set of differential equations of the following form:

$$\dot{\mathbf{x}}(t) = \mathbf{f}(\mathbf{x},t) + H\mathbf{x}(t) + B(\mathbf{x},t)\mathbf{u}(t) \tag{5.4}$$

$$\mathbf{y}(t) = \mathbf{C}^T\mathbf{x}(t) \tag{5.5}$$

The model described by Equations 5.4 and 5.5 has been used for the development of TDC. Additionally, uncertain disturbances may affect the system at any instant of time during operation. It is possible to extrapolate the operating modes of a bioreactor system from three (batch, continuous, and fed batch) to nine [15]. This is shown in Table 3.1.

5.3 Time delay control

TDC is a recent development that neither requires estimation of parameters of the system nor is dependent on training, as in artificial neural net (ANN)–based controllers. It depends on estimation of a function representing the effect of uncertainties. This is done by assuming that the value of the function representing the effect of uncertainties remains almost the same over an interval.

The TDC law is derived assuming that the plant states follow a certain prefixed reference trajectory. The control objective is to force the error (difference between the plant and model state vectors) to zero. These types of controllers are basically robust in nature. A block diagram of a time delay controller is shown in Figure 5.1. The condition to be satisfied for the application of time delay controller is that all the states along with their derivatives are available.

FIGURE 5.1 Schematic block diagram of time delay controller.

Adding the uncertainty factor, the plant model can be represented by modifying Equation 5.4 as

$$\dot{\mathbf{x}}(t) = \mathbf{f}(\mathbf{x},t) + H(\mathbf{x}(t) + B(\mathbf{x},t)\mathbf{u}(t) + \mathbf{h}(\mathbf{x},t) + \mathbf{d}(t) \qquad (5.6)$$

where
  $\mathbf{h}(\mathbf{x},t)$ is a four-element nonlinear vector representing the unmodeled part of the plant dynamics.
  $\mathbf{d}(t)$ is an unknown disturbance vector of four dimensions.

The linear time invariant (LTI) model that generates the desired trajectory takes the form of

$$\dot{\mathbf{x}}_\mathbf{d}(t) = A_m \mathbf{x}_\mathbf{d}(t) + B_m \mathbf{r}(t) \qquad (5.7)$$

where
  $\mathbf{x}_\mathbf{d}(t)$ is the four-dimensional model state vector.
  $\mathbf{r}(t)$ is a four-element command vector.
  A_m is a (4×4) constant matrix.

The error vector $\mathbf{e}(t)$ is defined as

$$\mathbf{e}(t) = \mathbf{x}_\mathbf{d}(t) - \mathbf{x}(t)$$

The control objective is to force the error vector to vanish with dynamics given by

$$\dot{\mathbf{e}}(t) = A_m \mathbf{e}(t) \qquad (5.8)$$

A_m is selected such that its diagonal elements are the highest slopes of the optimal trajectories obtained using an "observer" system. Off-diagonal elements of the A_m matrix are chosen arbitrarily, and are at least an order less than the diagonal values.
From Equation 5.8,

$$\dot{\mathbf{x}}_\mathbf{d}(t) - \dot{\mathbf{x}}(t) = A_m(\mathbf{x}_\mathbf{d}(t) - \mathbf{x}(t))$$

or

$$\dot{\mathbf{x}}_{\mathbf{d}}(t) - A_m\mathbf{x}_{\mathbf{d}}(t) = \dot{\mathbf{x}}(t) - A_m\mathbf{x}(t) \qquad (5.9)$$

Note that in bioprocesses a reference model in the form of Equation 5.7 does not exist. This forces time delay controller to follow the optimal trajectory defined earlier.

However, Equation 5.9 can help in getting a reference model of the form

$$\dot{\mathbf{x}}_{\mathbf{d}}(t) - A_m\mathbf{x}_{\mathbf{d}}(t) \cong \mathbf{r}(t) \qquad (5.10)$$

The error equation can be written in the following form:

$$\dot{\mathbf{e}}(t) = A_m\mathbf{x}_{\mathbf{d}}(t) + B_m\mathbf{r}(t) - \mathbf{f}(\mathbf{x},t) - H\mathbf{x}(t)$$
$$- B(\mathbf{x},t)\mathbf{u}(t) - \mathbf{h}(\mathbf{x},t) - \mathbf{d}(t)$$

or

$$\dot{\mathbf{e}}(t) = A_m\mathbf{e}(t) + A_m\mathbf{x}(t) + B_m\mathbf{r}(t) - \mathbf{f}(\mathbf{x},t)$$
$$- H\mathbf{x}(t) - B(\mathbf{x},t)\mathbf{u}(t) - \mathbf{h}(\mathbf{x},t) - \mathbf{d}(t) \qquad (5.11)$$

Comparing Equations 5.8 and 5.11, the following expression is obtained:

$$A_m\mathbf{x}(t) + B_m\mathbf{r}(t) - \mathbf{f}(\mathbf{x},t) - H\mathbf{x}(t)$$
$$- B(\mathbf{x},t)\mathbf{u}(t) - \mathbf{h}(\mathbf{x},t) - \mathbf{d}(t) = 0 \qquad (5.12)$$

Equation 5.12 cannot always be satisfied because the number of controls is generally not equal to the number of states. Thus an approximate solution of Equation 5.12 is adopted as follows:

$$\mathbf{u}(t) = B^+(\mathbf{x},t)[-\mathbf{f}(\mathbf{x},t) - H\mathbf{x}(t)$$
$$- \mathbf{h}(\mathbf{x},t) - \mathbf{d}(t) + A_m\mathbf{x}(t) + B_m\mathbf{r}(t)] \qquad (5.13)$$

where $B^+(\mathbf{x},t)$ is the matrix pseudoinverse of $B(\mathbf{x},t)$ and is defined as $B^+(\mathbf{x},t) = (B^T(\mathbf{x},t)B(\mathbf{x},t))^{-1}B^T(\mathbf{x},t)$. The condition for which Equation 5.13 exactly satisfies Equation 5.12 is determined as follows.

Substituting Equation 5.13 in Equation 5.6, the following is obtained:

$$\dot{\mathbf{x}}(t) = f(\mathbf{x},t) + H\mathbf{x}(t) + h(\mathbf{x},t) + \mathbf{d}(t) + B(\mathbf{x},t)B^+(\mathbf{x},t)[-f(\mathbf{x},t)$$
$$-H\mathbf{x}(t) - h(\mathbf{x},t) - \mathbf{d}(t) + A_m\mathbf{x}(t) + B_m\mathbf{r}(t)]$$

Again from Equation 5.8, it follows that

$$\dot{\mathbf{e}}(t) = A_m\mathbf{e}(t) + [I - B(\mathbf{x},t)B^+(\mathbf{x},t)][-f(\mathbf{x},t)$$
$$- H\mathbf{x}(t) - h(\mathbf{x},t) - \mathbf{d}(t) + A_m\mathbf{x}(t) + B_m\mathbf{r}(t)] \quad (5.14)$$

In order to obtain the desired dynamics given by Equation 5.8, the following structural constraint is to be fulfilled:

$$[I - B(\mathbf{x},t)B^+(\mathbf{x},t)][-\mathbf{f}(\mathbf{x},t) - H\mathbf{x}(t) - h(\mathbf{x},t)$$
$$- \mathbf{d}(t) + A_m\mathbf{x}(t) + B_m\mathbf{r}(t)] = 0 \quad (5.15)$$

If $B(\mathbf{x},t)$ is a square matrix of full rank, the above structural constraint is easily satisfied. If it is not so, the choice of the reference model is somewhat restricted. Moreover, some elements of the unknown dynamics vector $h(\mathbf{x},t)$ and unexpected disturbance vector $\mathbf{d}(t)$ should be known in order to ensure that the system satisfies the above constraint. It can be shown that this condition is always satisfied for systems expressed in canonical form.

It is of interest to determine the control action $\mathbf{u}(t)$ that will force the plant to follow the reference model in the face of unknown dynamics $h(\mathbf{x},t)$ and unexpected disturbance $\mathbf{d}(t)$. These two terms can be determined from the plant dynamic in Equation 5.6:

$$h(\mathbf{x},t) + \mathbf{d}(t) = \dot{\mathbf{x}}(t) - \mathbf{f}(\mathbf{x},t) - H\mathbf{x}(t) - B(\mathbf{x},t)\mathbf{u}(t) \quad (5.16)$$

In order to estimate the effect of the term $(h(\mathbf{x},t) + \mathbf{d}(t))$, it is considered in time delay controller that the value of the function $(h(\mathbf{x},t) + \mathbf{d}(t))$ at the present time t is very close to that at time $(t - L)$ in the past for a small time delay L, that is,

$$h(\mathbf{x},t) + \mathbf{d}(t) \cong h(\mathbf{x},(t - L)) + \mathbf{d}(t - L) \quad (5.17)$$

Combining Equations 5.16 and 5.17, the effect of the function $h(\mathbf{x},t) + \mathbf{d}(t)$ is estimated by

$$h(\mathbf{x},t) + \mathbf{d}(t) \cong \dot{\mathbf{x}}(t - L) - \mathbf{f}(\mathbf{x},t - L)$$
$$- H\dot{\mathbf{x}}(t - L) - B(\mathbf{x},t - L)\mathbf{u}(t - L) \quad (5.18)$$

It is important to estimate the upper limit of L such that Equation 5.17 approximately holds.

The TDC law is obtained by substituting Equation 5.18 in Equation 5.13 and is given by

$$
\begin{aligned}
\mathbf{u}(t) = B^+(\mathbf{x},t)[&-\mathbf{f}(\mathbf{x},t) - H\mathbf{x}(t) - \dot{\mathbf{x}}(t - L) \\
&+ \mathbf{f}(\mathbf{x},t - L) + H\mathbf{x}(t - L) + B(\mathbf{x},t - L)\mathbf{u}(t - L) \\
&+ A_m\mathbf{x}(t) + B_m\mathbf{r}(t)]
\end{aligned}
\tag{5.19}
$$

In the above equations the term $[-\mathbf{f}(\mathbf{x},t) - H\mathbf{x}(t) - \dot{\mathbf{x}}(t - L) + \mathbf{f}(\mathbf{x},t - L) + H\mathbf{x}(t - L) + B(\mathbf{x},t - L)\mathbf{u}(t - L)]$ attempts to cancel the nonlinear dynamics $\mathbf{f}(\mathbf{x},t)$, unknown dynamics $\mathbf{h}(\mathbf{x},t)$, and unexpected disturbances $d(t)$; the term $A_m\mathbf{x}(t) + B_m\mathbf{r}(t)$ inserts the desired dynamics of the reference model. Thus this controller observes the states and inputs of the system at a time $(t - L)$ and determines the control action at time t and hence is known as the time delay controller.

5.4 Time delay controller as bioprocess controller

In the case of the bioprocess, the reference model is represented as in Equation 5.10, that is,

$$
\dot{\mathbf{x}}_{\mathbf{d}}(t) - A_m\mathbf{x}_d(t) = B_m\mathbf{r}(t)
$$

where $\mathbf{x}_d(t)$ and $\dot{\mathbf{x}}_{\mathbf{d}}(t)$ represent the reference state trajectory and its time differentiation, respectively. These are obtained (as discussed earlier) by applying genetic algorithms [16] to model equations of the bioprocess system (Section 3.1). The command input $B_m\mathbf{r}(t)$ is calculated at each sampling instant using the above, which in turn is used to calculate the control input.

Substituting, the TDC law as applicable to the bioprocess is

$$
\begin{aligned}
\mathbf{u}(t) = B^+(\mathbf{x},t)[&-\mathbf{f}(\mathbf{x},t) - H\mathbf{x}(t) - \dot{\mathbf{x}}(t - L) + \mathbf{f}(\mathbf{x},t - L) \\
&+ H\mathbf{x}(t - L) + B(\mathbf{x},t - L)\mathbf{u}(t - L) \\
&+ \dot{\mathbf{x}}_{\mathbf{d}}(t) - A_m(\mathbf{x}_d(t) - \mathbf{x}(t))]
\end{aligned}
\tag{5.20}
$$

Let

$$
A_m = \begin{bmatrix} a_{11} & a_{12} & a_{13} & a_{14} \\ a_{21} & a_{22} & a_{23} & a_{24} \\ a_{31} & a_{32} & a_{33} & a_{34} \\ a_{41} & a_{42} & a_{43} & a_{44} \end{bmatrix}
$$

A_m is adapted as below:

$$
\begin{bmatrix}
(a_{11} - k_9\ error\ 1) & a_{12} & a_{13} \\
a_{21} & (a_{22} - k_9\ error\ 2) & a_{23} \\
a_{31} & a_{32} & (a_{33} - k_9\ error\ 3) \\
a_{41} & a_{42} & a_{43}
\end{bmatrix}
$$

$$
\begin{bmatrix}
a_{14} \\
a_{24} \\
a_{34} \\
(a_{44} - k_9\ error\ 4)
\end{bmatrix}
$$

where

$error\ 1 = (x_{d1}(t) - z_1(t))$
$error\ 2 = (x_{d2}(t) - z_2(t))$
$error\ 3 = (x_{d3}(t) - z_3(t))$
$error\ 4 = (x_{d4}(t) - z_4(t))$

z_1, z_2, z_3, and z_4 represent the estimated values of the states and k_9 is heuristically selected.

The flowchart (Figure 5.2) shows the functioning of the time delay controller. While implementing the TDC law, it was found that elements of corresponding values of $B(x,t)$ are very small. To avoid numerical difficulties, the following algebraic manipulation was done. Let

$$B_{new} = kB(\mathbf{x},t),\ \text{where}\ k = 10^4$$

$$D_{new} = B_{new}^+ = k^{-1}[B(\mathbf{x},t)]^+;$$
$$B_{old} = B(\mathbf{x},t - L);$$
$$T = D_{new} * B_{old}$$

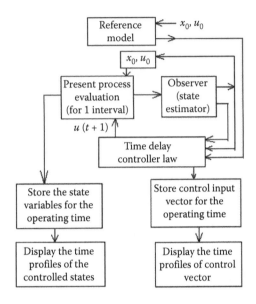

FIGURE 5.2 Flowchart of time delay controller applied to the bioprocess.

Now, the modified TDC law takes the form

$$\mathbf{u}_1(t) = [T]\mathbf{u}(t - L) + D_{new}[-\mathbf{f}(\mathbf{x},t) - H\mathbf{x}(t) - \dot{\mathbf{x}}(t - L) \\ + \mathbf{f}(\mathbf{x}, t - L) + H\mathbf{x}(t - L) + \dot{\mathbf{x}}_{\mathbf{d}}(t) - A_m(\mathbf{x}_{\mathbf{d}}(t) - \mathbf{x}(t))]$$

or

$$k\mathbf{u}_1(t) = [B(\mathbf{x},t)]^+ B(\mathbf{x}, t - L)\mathbf{u}(t - L) \\ + [B(\mathbf{x},t)]^+[-\mathbf{f}(\mathbf{x},t) - H\mathbf{x}(t) - \dot{\mathbf{x}}(t - L) \\ + \mathbf{f}(\mathbf{x}, t - L) + H\mathbf{x}(t - L) + \dot{\mathbf{x}}_{\mathbf{d}}(t) - A_m(\mathbf{x}_{\mathbf{d}}(t) - \mathbf{x}(t))]$$

$$(5.21)$$

Comparing Equations 5.20 and 5.21, the TDC law becomes

$$\mathbf{u}(t) = k\mathbf{u}_1(t) \tag{5.22}$$

where

$$\mathbf{u}_1(t) = [F_{01}(t) F_{11}(t) F_{41}(t) (K_L a_1(t))]^T$$

The control inputs are calculated as

$$F_0(t) = kF_{01}(t) \tag{5.23}$$

$$F_1(t) = kF_{11}(t) \tag{5.24}$$

$$F_4(t) = kF_{41}(t) \tag{5.25}$$

$$K_L a(t) = k(K_L a_1(t)) \tag{5.26}$$

where the expressions for $F_{01}(t)$, $F_{11}(t)$, $F_{41}(t)$, and $K_L a_1(t)$ are obtained as follows:

$$
\begin{aligned}
F_{01}(t) = &\, (t_{11}F_0(t-L) + t_{12}F_1(t-L) \\
&+ t_{13}F_4(t-L) + t_{14}K_L a(t-L)) \\
&+ d_{11}(f_1(\mathbf{x}, t-L) - f_1(\mathbf{x}, t) + \dot{x}_{d1}(t) - \dot{x}_1(t-L)) \\
&+ d_{12}(f_2(\mathbf{x}, t-L) - f_2(\mathbf{x}, t) + \dot{x}_{d2}(t) - \dot{x}_2(t-L)) \\
&+ d_{13}(f_3(\mathbf{x}, t-L) - f_3(\mathbf{x}, t) + \dot{x}_{d3}(t) - \dot{x}_3(t-L)) \\
&+ d_{14}(f_4(\mathbf{x}, t-L) - f_4(\mathbf{x}, t) + \dot{x}_{d4}(t) - \dot{x}_4(t-L)) \\
&+ (d_{11}(-k_d(x_1(t-L) - x_1(t)) \\
&+ d_{12}\left(-\left(m_s + \frac{\alpha}{y_p}\right)(x_1(t-L) - x_1(t))\right) \\
&+ d_{13}\left(-\left(m_{s0} + \frac{\alpha}{y_{p0}}\right)(x_1(t-L) - x_1(t))\right) \\
&+ d_{14}(\alpha(x_1(t-L) - x_1(t))) \\
&+ d_{11}(a_{11}(x_1(t) - x_{d1}(t)) + a_{12}(x_2(t) \\
&- x_{d2}(t)) + a_{13}(x_3(t) - x_{d3}(t)) \\
&+ a_{14}(x_4)(t) - x_{d4}(t))) + d_{12}(a_{21}(x_1(t) - x_{d1}(t)) \\
&+ a_{22}(x_2(t) - x_{d2}(t)) \\
&+ a_{23}(x_3(t) - x_{d3}(t)) + a_{24}(x_4(t) - x_{d4}(t))) \\
&+ d_{13}(a_{31}(x_1(t) - x_{d1}(t)) \\
&+ a_{32}(x_2(t) - x_{d2}(t)) + a_{33}(x_3(t) - x_{d3}(t)) \\
&+ a_{34}(x_4(t) - x_{d4}(t))) \\
&+ d_{14}(a_{41}(x_1(t) - x_{d1}(t)) + a_{42}(x_2(t) - x_{d2}(t)) \\
&+ a_{43}(x_3(t) - x_{d3}(t)) \\
&+ a_{44}(x_4(t) - x_{d4}(t)))
\end{aligned}
$$

$$F_{11}(t) = (t_{21}F_0(t-L) + t_{22}F_1(t-L)$$
$$+ t_{23}F_4(t-L) + t_{24}K_La(t-L))$$
$$+ d_{21}(f_1(\mathbf{x}, t-L) - f_1(\mathbf{x}, t) + \dot{x}_{d1}(t) - \dot{x}_1(t-L))$$
$$+ d_{22}(f_2(\mathbf{x}, t-L) - f_2(\mathbf{x}, t) + \dot{x}_{d2}(t) - \dot{x}_2(t-L))$$
$$+ d_{23}(f_3(\mathbf{x}, t-L) - f_3(\mathbf{x}, t) + \dot{x}_{d3}(t) - \dot{x}_3(t-L))$$
$$+ d_{24}(f_4(\mathbf{x}, t-L) - f_4(\mathbf{x}, t) + \dot{x}_{d4}(t) - \dot{x}_4(t-L))$$
$$+ (d_{21}(-k_d(x_1(t-L) - x_1(t)))$$
$$+ d_{22}\left(-\left(m_s + \frac{\alpha}{y_p}\right)(x_1(t-L) - x_1(t))\right)$$
$$+ d_{23}\left(-\left(m_{s0} + \frac{\alpha}{y_{p0}}\right)(x_1(t-L) - x_1(t))\right)$$
$$+ d_{24}(\alpha(x_1(t-L) - x_1(t))))$$
$$+ d_{21}(a_{11}(x_1(t) - x_{d1}(t)) + a_{12}(x_2(t) - x_{d2}(t))$$
$$+ a_{13}(x_3(t) - x_{d3}(t))$$
$$+ a_{14}(x_4)(t) - x_{d4}(t))) + d_{22}(a_{21}(x_1(t) - x_{d1}(t))$$
$$+ a_{22}(x_2(t) - x_{d2}(t))$$
$$+ a_{23}(x_3(t) - x_{d3}(t)) + a_{24}(x_4(t) - x_{d4}(t)))$$
$$+ d_{23}(a_{31}(x_1(t) - x_{d1}(t))$$
$$+ a_{32}(x_2(t) - x_{d2}(t)) + a_{33}(x_3(t) - x_{d3}(t))$$
$$+ a_{34}(x_4(t) - x_{d4}(t)))$$
$$+ d_{24}(a_{41}(x_1(t) - x_{d1}(t)) + a_{42}(x_2(t) - x_{d2}(t))$$
$$+ a_{43}(x_3(t) - x_{d3}(t))$$
$$+ a_{44}(x_4(t) - x_{d4}(t)))$$

$$F_{41}(t) = (t_{31}F_0(t-L) + t_{32}F_1(t-L)$$
$$+ t_{33}F_4(t-L) + t_{34}K_La(t-L))$$
$$+ d_{31}(f_1(\mathbf{x}, t-L) - f_1(\mathbf{x}, t) + \dot{x}_{d1}(t) - \dot{x}_1(t-L))$$
$$+ d_{32}(f_2(\mathbf{x}, t-L) - f_2(\mathbf{x}, t) + \dot{x}_{d2}(t) - \dot{x}_2(t-L))$$
$$+ d_{33}(f_3(\mathbf{x}, t-L) - f_3(\mathbf{x}, t) + \dot{x}_{d3}(t) - \dot{x}_3(t-L))$$
$$+ d_{34}(f_4(\mathbf{x}, t-L) - f_4(\mathbf{x}, t) + \dot{x}_{d4}(t) - \dot{x}_4(t-L))$$
$$+ (d_{31}(-k_d(x_1(t-L) - x_1(t)) + d_{32}$$
$$\left(-\left(m_s + \frac{\alpha}{y_p}\right)(x_1(t-L) - x_1(t))\right)$$

$$+ d_{33}\left(-\left(m_{s0} + \frac{\alpha}{y_{p0}}\right)(x_1(t-L) - x_1(t))\right)$$
$$+ d_{34}(\alpha(x_1(t-L) - x_1(t))))$$
$$+ d_{31}(a_{11}(x_1(t) - x_{d1}(t)) + a_{12}(x_2(t) - x_{d2}(t))$$
$$+ a_{13}(x_3(t) - x_{d3}(t))$$
$$+ a_{14}(x_4(t) - x_{d4}(t))) + d_{32}(a_{21}(x_1(t)$$
$$- x_{d1}(t)) + a_{22}(x_2(t) - x_{d2}(t))$$
$$+ a_{23}(x_3(t) - x_{d3}(t)) + a_{24}(x_4(t) - x_{d4}(t)))$$
$$+ d_{33}(a_{31}(x_1(t) - x_{d1}(t))$$
$$+ a_{32}(x_2(t) - x_{d2}(t)) + a_{33}(x_3(t) - x_{d3}(t))$$
$$+ a_{34}(x_4(t) - x_{d4}(t)))$$
$$+ d_{34}(a_{41}(x_1(t) - x_{d1}(t)) + a_{42}(x_2(t)$$
$$- x_{d2}(t)) + a_{43}(x_3(t) - x_{d3}(t))$$
$$+ a_{44}(x_4(t) - x_{d4}(t)))$$

$$K_L a1(t) = (t_{41}F_0(t-L) + t_{42}F_1(t-L)$$
$$+ t_{43}F_4(t-L) + t_{44}K_L a(t-L))$$
$$+ d_{41}(f_1(\mathbf{x}, t-L) - f_1(\mathbf{x}, t) + \dot{x}_{d1}(t) - \dot{x}_1(t-L))$$
$$+ d_{42}(f_2(\mathbf{x}, t-L) - f_2(\mathbf{x}, t) + \dot{x}_{d2}(t) - \dot{x}_2(t-L))$$
$$+ d_{43}(f_3(\mathbf{x}, t-L) - f_3(\mathbf{x}, t) + \dot{x}_{d3}(t) - \dot{x}_3(t-L))$$
$$+ d_{44}(f_4(\mathbf{x}, t-L) - f_4(\mathbf{x}, t) + \dot{x}_{d4}(t) - \dot{x}_4(t-L))$$
$$+ (d_{41}(-k_d(x_1(t-L) - x_1(t))$$
$$+ d_{42}\left(-\left(m_s + \frac{\alpha}{y_p}\right)(x_1(t-L) - x_1(t))\right)$$
$$+ d_{43}\left(-\left(m_{s0} + \frac{\alpha}{y_{p0}}\right)(x_1(t-L) - x_1(t))\right)$$
$$+ d_{44}(\alpha(x_1(t-L) - x_1(t))))$$
$$+ d_{41}(a_{11}(x_1(t) - x_{d1}(t)) + a_{12}(x_2(t)$$
$$- x_{d2}(t)) + a_{13}(x_3(t) - x_{d3}(t))$$
$$+ a_{14}(x_4(t) - x_{d4}(t))) + d_{42}(a_{21}(x_1(t)$$
$$- x_{d1}(t)) + a_{22}(x_2(t) - x_{d2}(t))$$
$$+ a_{23}(x_3(t) - x_{d3}(t)) + a_{24}(x_4(t) - x_{d4}(t)))$$
$$+ d_{43}(a_{31}(x_1(t) - x_{d1}(t))$$

$$+ a_{32}(x_2(t) - x_{d2}(t)) + a_{33}(x_3(t) - x_{d3}(t))$$
$$+ a_{34}(x_4(t) - x_{d4}(t)))$$
$$+ d_{44}(a_{41}(x_1(t) - x_{d1}(t)) + a_{42}(x_2(t) - x_{d2}(t))$$
$$+ a_{43}(x_3(t) - x_{d3}(t))$$
$$+ a_{44}(x_4(t) - x_{d4}(t)))$$

As per Figure 5.2, reference model generates the reference trajectories. The initial conditions are updated every time with the final values of the last iterations. Equations 5.23 through 5.26 describe the control action at the end of each sampling instant. Fairly good results have been achieved even while inserting some upper and lower bounds to the control inputs. These bounds are generally dictated by the capacity of the pump, maximum speed of the stirrer, and so forth.

5.5   Observer design

It was assumed in deriving the TDC law that all the states along with their derivatives are available. In the present process, only dissolved oxygen is a measurable output of the process. The other states are to be estimated. A suitable observer is to be designed to estimate the unmeasurable states.

Designing an observer for a nonlinear system is fairly complicated, and the working of a nonlinear observer may offset the advantage gained by the simple TDC law. Since TDC forces the plant dynamics to follow the reference model, which is an LTI model, a strategy for the observer design is to use the LTI reference model instead of the actual plant model. A Luenberger full-order observer has been designed.

The observer equation is

$$\dot{\mathbf{z}}(t) = A_m \mathbf{z}(t) + B_m \mathbf{r}(t) + \mathbf{fc}^T(\mathbf{z}(t) - \mathbf{x}(t))$$

Substituting from Equation 5.7,

$$\dot{\mathbf{z}}(t) = A_m \mathbf{z}(t) + \dot{\mathbf{x}}_d(t) - A_m \mathbf{x}_d(t) + \mathbf{fc}^T(\mathbf{z}(t) - \mathbf{x}(t))$$

or

$$\dot{\mathbf{z}}(t) = \dot{\mathbf{x}}_d(t) + A_m(\mathbf{z}(t) - \mathbf{x}(t)) + \mathbf{fc}^T(\mathbf{z}(t) - \mathbf{x}(t)) \qquad (5.27)$$

or, explicitly,

$$\dot{z}_1(t) = \dot{x}_{d1}(t) - (a_{11}(x_{d1}(t) - z_1(t)) + a_{12}(x_{d2}(t)$$
$$- z_2(t)) + a_{13}(x_{d3}(t) - z_3(t))$$
$$+ a_{14}(x_{d4}(t) - z_4(t))) + f_1(z_3(t) - x_3(t)) \qquad (5.28)$$

$$\dot{z}_2(t) = \dot{x}_{d2}(t) - (a_{21}(x_{d1}(t) - z_1(t)) + a_{22}(x_{d2}(t)$$
$$- z_2(t)) + a_{23}(x_{d3}(t) - z_3(t))$$
$$+ a_{24}(x_{d4}(t) - z_4(t))) + f_2(z_3(t) - x_3(t)) \qquad (5.29)$$

$$\dot{z}_3(t) = \dot{x}_{d3}(t) - (a_{31}(x_{d1}(t) - z_1(t)) + a_{32}(x_{d2}(t)$$
$$- z_2(t)) + a_{33}(x_{d3}(t) - z_3(t))$$
$$+ a_{34}(x_{d4}(t) - z_4(t))) + f_3(z_3(t) - x_3(t)) \qquad (5.30)$$

$$\dot{z}_4(t) = \dot{x}_{d4}(t) - (a_{41}(x_{d1}(t) - z_1(t)) + a_{42}(x_{d2}(t)$$
$$- z_2(t)) + a_{43}(x_{d3}(t) - z_3(t))$$
$$+ a_{44}(x_{d4}(t) - z_4(t))) + f_4(z_3(t) - x_3(t)) \qquad (5.31)$$

By solving Equations 5.28 through 5.31, unmeasurable states can be estimated using input and output data. The gain vector $\mathbf{f}$ is found by using the standard algorithms of Ackerman and Gura for gain vector calculations [48].

TDC law with the estimated states takes the form

$$\mathbf{u}(t) = B^+(\mathbf{z},t)[-f(\mathbf{z},t) - H\mathbf{z}(t) - \dot{\mathbf{z}}(t - L) + f(\mathbf{z},t - L)$$
$$+ B(\mathbf{z},t - L)\mathbf{u}(t - L)$$
$$+ H\mathbf{z}(t - L) + A_m\mathbf{z}(t) + \dot{\mathbf{x}}_d(t) - A_m\mathbf{x}_d(t)] \qquad (5.32)$$

To ascertain stability of the overall system, that is, the plant with the controller and observer, the observer gain vector $\mathbf{f}$ is substituted in the following expression [43,44]:

$$det\{(\lambda I - A_m) - B(\mathbf{x},t)B^+(\mathbf{x},t)(e^{-L\lambda}\lambda I - A_m)$$
$$(\lambda I - A_0)^{-1}(\lambda I - A_m)\} = 0$$

where $A_0 = A_m + \mathbf{f}\mathbf{c}^T$; the resultant eigenvalues should lie in the closed left half plane. A schematic block diagram of a time delay controller with the proposed observer is shown in Figure 5.3.

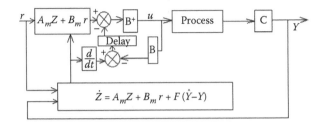

FIGURE 5.3 Schematic block diagram of the observer with a time delay controller.

5.6 PID controllers for bioprocess control

By using a PID controller, it is possible to control the dissolved oxygen concentration in the bioreactor. Controller parameters k_1, k_2, and k_3 are fixed beforehand. In this chapter, a PID controller has also been used for tracking oxygen concentration.

However, the control input ($K_L a$) is bounded by physical constraints.

The controller is given by

$$m(t) = K_p \left[e(t) + \frac{1}{T_i} \int e(t)dt + T_d \frac{de(t)}{dt} \right] \qquad (5.33)$$

where $m(t)$ is the control signal generated and $e(t)$ is the error in the tracking oxygen concentration.

In the discrete time domain, the above equation is written as

$$m(k) = K_p e(k) + \frac{K_p}{T_i} \sum_{i=1}^{k} e(i)\Delta t$$
$$+ T_d K_p \frac{(e(k) - e(k-1))}{\Delta t} \qquad (5.34)$$

Δt is the sampling interval.

In the next instant the control input is

$$m(k+1) = K_p e(k+1) + \frac{K_p}{T_i} \sum_{i=1}^{k+1} e(i)\Delta t$$
$$+ T_d K_p \frac{(e(k+1) - e(k))}{\Delta t} \qquad (5.35)$$

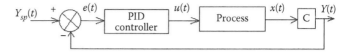

FIGURE 5.4   Schematic block diagram of a PID controller.

Subtracting Equation 5.34 from Equation 5.35,

$$m(k + 1) = m(k) + k_1 e(k + 1) + k_2 e(k) + k_3 e(k - 1) \quad (5.36)$$

with $k_1 = k_p[1 + (\Delta T/T_i) + (T_i/\Delta t)]$, $k_2 = -k_p[1 + (2T_d/\Delta t)]$,   and $k_3 = K_p(T_d/\Delta t)$.

The schematic block diagram of a PID controller is shown in Figure 5.4.

5.7 Simulation results and discussions

Results of TDC-controlled states along with error profiles (with respect to reference trajectories) and control input trajectories are shown in Figures 5.5 through 5.31 for op-modes 1–9. The figures are self-explanatory. It is generally seen that the performance of the time delay controller is fair as a bioprocess controller.

In a separate set of experimentations, comparisons of the performance of the time delay controller and PID controllers (which regulate the dissolved oxygen concentration to the pre-defined optimal value) for operating modes 2–9 have been carried out. The following inferences have been drawn from the results of experimentations.

- Figure 5.32 represents time delay controller- and PID-controlled oxygen profiles along with error profiles with respect to the optimal oxygen concentration in op-mode 2. Comparing, it could be inferred that the PID controller works better than the time delay controller in oxygen concentration regulation in this mode.

- For op-mode 3, Figure 5.33 represents the TDC-controlled oxygen concentration and corresponding error profile and the PID-controlled oxygen concentration and error profile. Comparing, it can be inferred that oxygen is better controlled by the PID controller.

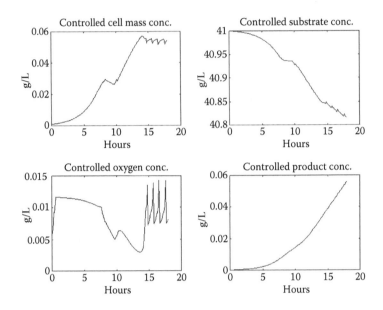

FIGURE 5.5 Time delay controller time profiles of single cell protein (SCP) fermentation process variables in op-mode 1.

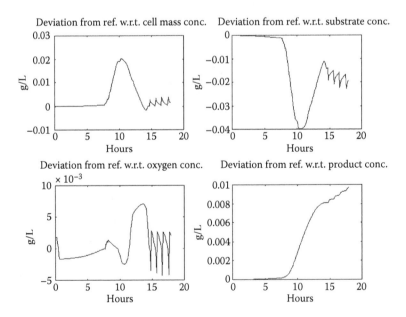

FIGURE 5.6 Deviation of time delay controller time profiles of SCP fermentation process variables in op-mode 1 from the reference time profiles.

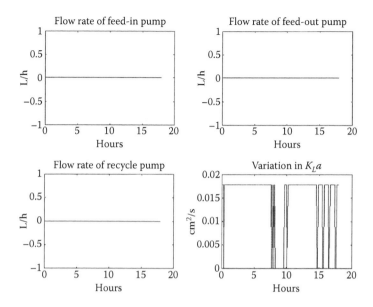

FIGURE 5.7 Time delay controller-generated control inputs for controlling SCP fermentation process in op-mode 1.

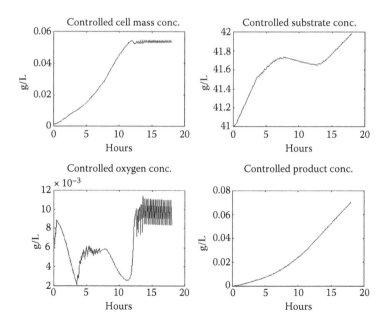

FIGURE 5.8 Time delay controller-controlled time profiles of SCP fermentation process in variables in op-mode 2.

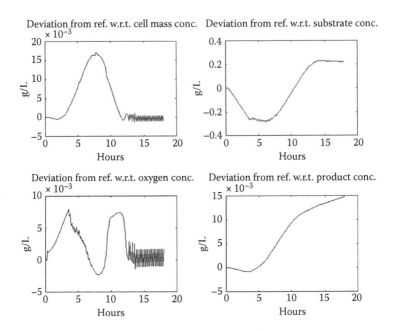

FIGURE 5.9 Deviations of time delay controller-controlled time profiles of SCP fermentation process variables in op-mode 2 from the reference time profiles.

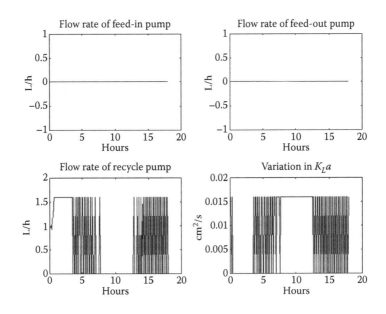

FIGURE 5.10 Time delay controller-generated control inputs for controlling SCP fermentation process in op-mode 2.

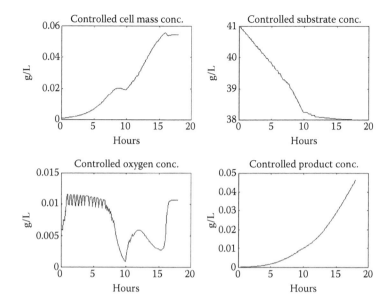

FIGURE 5.11 Time delay controller-controlled time profiles of SCP fermentation process variables in op-mode 3.

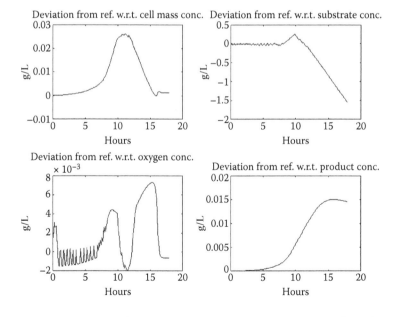

FIGURE 5.12 Deviations of time delay controller-controlled time profiles of SCP fermentation process variables in op-mode 3 from the reference time profiles.

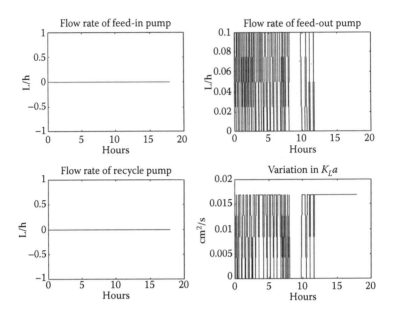

FIGURE 5.13 Time delay controller-generated control input for controlling SCP fermentation process in op-mode 3.

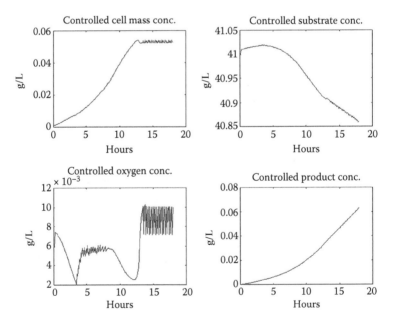

FIGURE 5.14 Time delay controller-controlled time profiles of SCP fermentation process variables in op-mode 4.

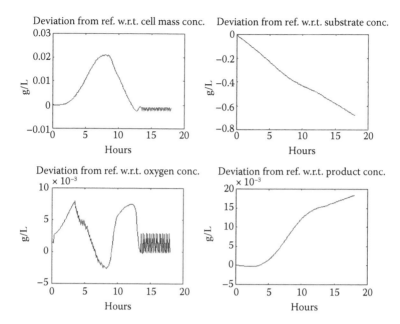

FIGURE 5.15 Deviations of time delay controller-controlled time profiles of SCP fermentation process variables in op-mode 4 from the reference time profiles.

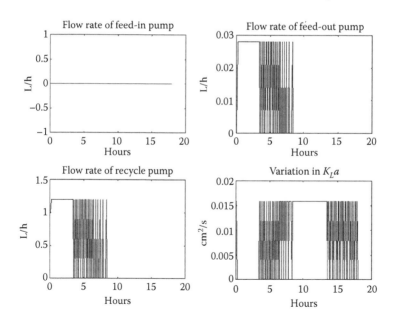

FIGURE 5.16 Time delay controller-generated control inputs for controlling SCP fermentation process in op-mode 4.

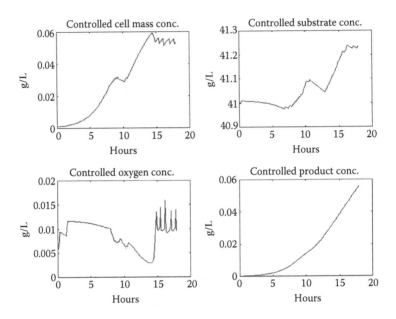

FIGURE 5.17 Time delay controller-controlled time profiles of SCP fermentation process variables in op-mode 5.

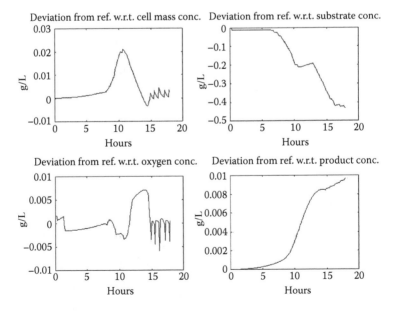

FIGURE 5.18 Deviations of time delay controller-controlled time profiles of SCP fermentation process variables in op-mode 5 from the reference time profiles.

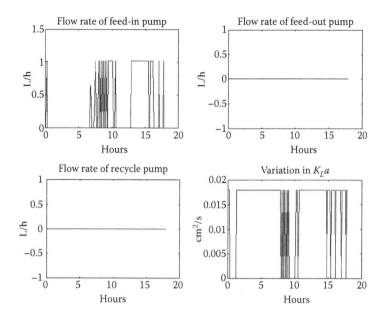

FIGURE 5.19 Time delay controller-generated control inputs for controlling SCP fermentation process in op-mode 5.

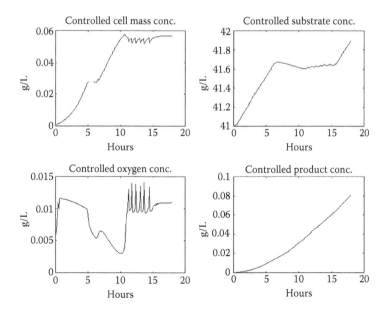

FIGURE 5.20 Time delay controller-controlled time profiles of SCP fermentation process variables in op-mode 6.

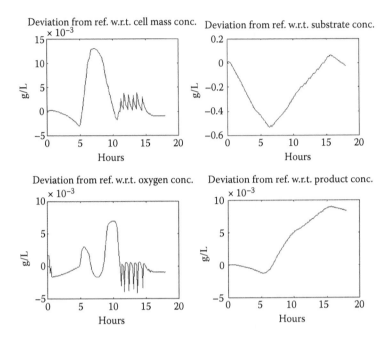

FIGURE 5.21 Deviations of time delay controller-controlled time profiles of SCP fermentation process variables in op-mode 6 from the reference time profiles.

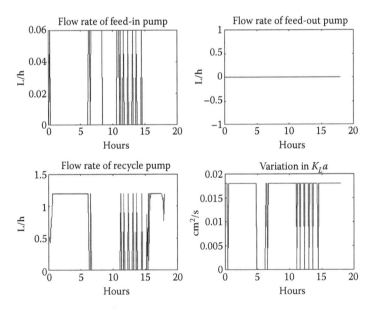

FIGURE 5.22 Time delay controller-generated control inputs for controlling SCP fermentation process in op-mode 6.

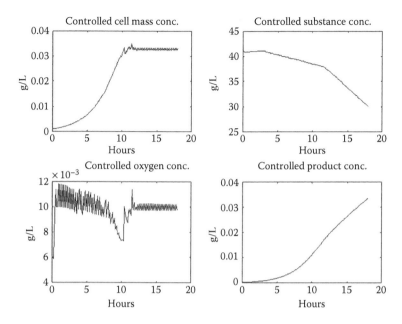

FIGURE 5.23 Time delay controller-controlled time profiles of SCP fermentation process variables in op-mode 7.

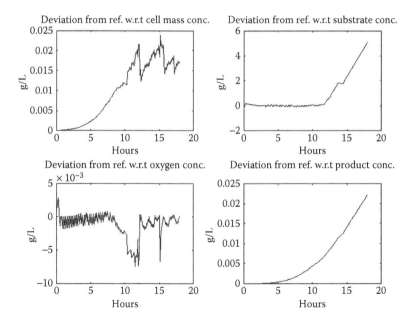

FIGURE 5.24 Deviations of time delay controller-controlled time profiles of SCP fermentation process variables in op-mode 7 from the reference time profiles.

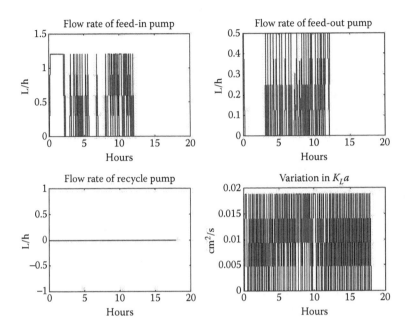

FIGURE 5.25 Time delay controller-generated control inputs for controlling SCP fermentation process in op-mode 7.

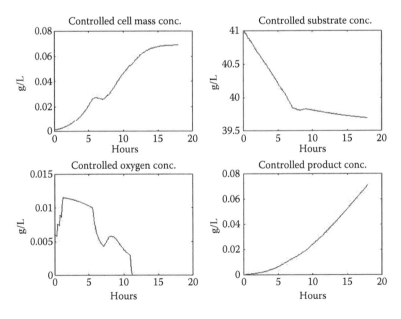

FIGURE 5.26 Time delay controller-controlled time profiles of SCP fermentation process variables in op-mode 8.

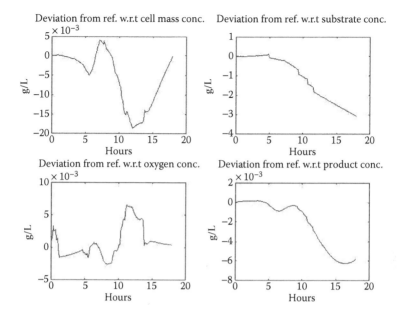

FIGURE 5.27 Deviations of time delay controller-controlled time profiles of SCP fermentation process variables in op-mode 8 from the reference time profiles.

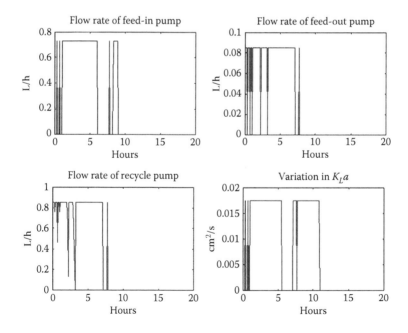

FIGURE 5.28 Time delay controller-generated control inputs for controlling SCP fermentation process in op-mode 8.

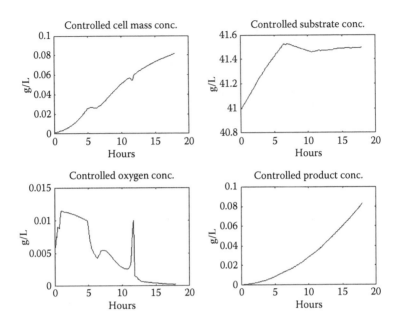

FIGURE 5.29 Time delay controller-controlled time profiles of SCP fermentation process variables in op-mode 9.

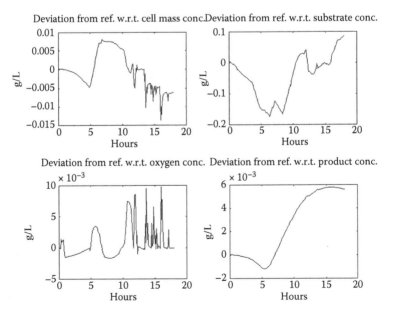

FIGURE 5.30 Deviations of time delay controller-controlled time profiles of SCP fermentation process variables in op-mode 9 from the reference time profiles.

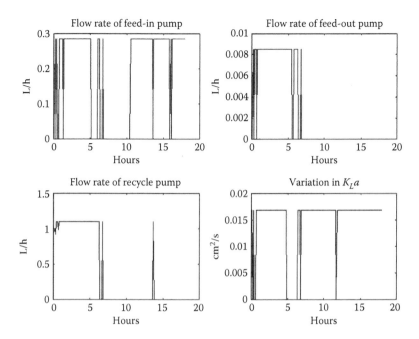

FIGURE 5.31 Time delay controller-generated control inputs for controlling SCP fermentation process in op-mode 9.

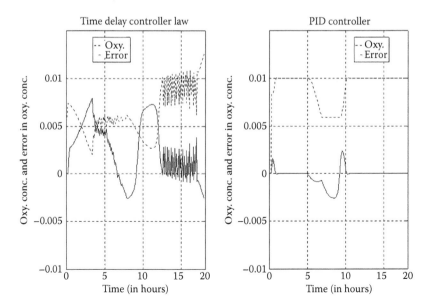

FIGURE 5.32 Comparison of time delay controller and PID for oxygen concentration control in op-mode 2.

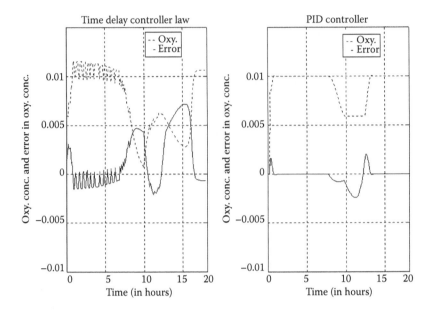

FIGURE 5.33 Comparison of time delay controller and PID for oxygen concentration control in op-mode 3.

- For op-mode 4, Figure 5.34 represents the time delay controller-controlled oxygen concentration and the corresponding error profile and also the PID-controlled oxygen concentration and error profile. Comparing the profiles, it can be inferred that in this mode oxygen is better controlled by the PID controller than they time delay controller.

- For op-mode 5, Figure 5.35 represents the time delay controller-controlled oxygen concentration and the corresponding error profile, and it also displays the PID-controlled oxygen concentration and its error profile. Comparing the error profiles of oxygen in Figure 5.35, it can be inferred that the PID controller controls oxygen concentration better than the time delay controller.

- For op-mode 6, Figure 5.36 represents the time delay controller-controlled oxygen concentration and the corresponding error profile and also the PID-controlled oxygen concentration and error profile. Comparing the error profiles of oxygen in Figure 5.36, it can be inferred that the PID control of oxygen concentration is better.

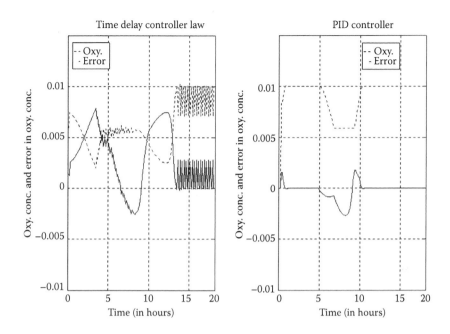

FIGURE 5.34 Comparison of time delay controller and PID for oxygen concentration control in op-mode 4.

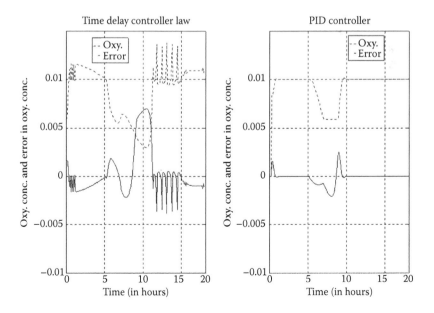

FIGURE 5.35 Comparison of time delay controller and PID for oxygen concentration control in op-mode 5.

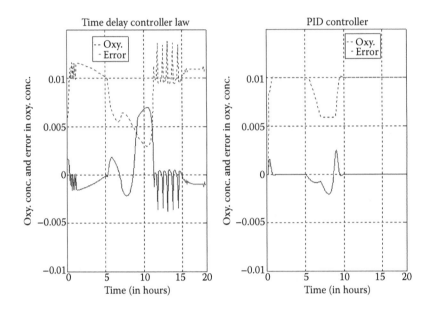

FIGURE 5.36 Comparison of time delay controller and PID for oxygen concentration control in op-mode 6.

- For op-mode 7, Figure 5.37 represents the time delay controller-controlled oxygen concentration and the corresponding error profile. It also displays the PID-controlled oxygen concentration and error profile. Comparing the error profiles of oxygen in Figure 5.37, it can be safely inferred that oxygen is better controlled by the PID controller.

- For op-mode 8, Figure 5.38 displays the time delay controller-controlled oxygen concentration and the corresponding error profiles. It also displays the PID-controlled oxygen concentration and error profile. Comparing the error profiles, it is seen that the PID controller has a shade better than the time delay controller with respect to oxygen concentration regulation.

- For op-mode 9, Figure 5.39 represents the time delay controller-controlled oxygen concentration and the corresponding error profile. It also represents the PID-controlled oxygen concentration and error profile. Comparing the two error profiles, it can be inferred that the time delay controller controls the oxygen concentration better than the PID controller.

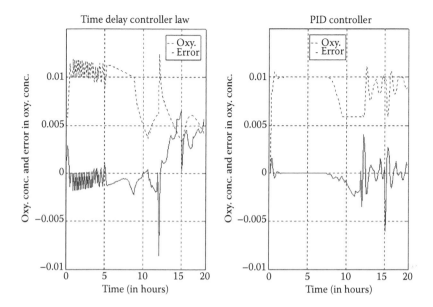

FIGURE 5.37 Comparison of time delay controller and PID for oxygen concentration control in op-mode 7.

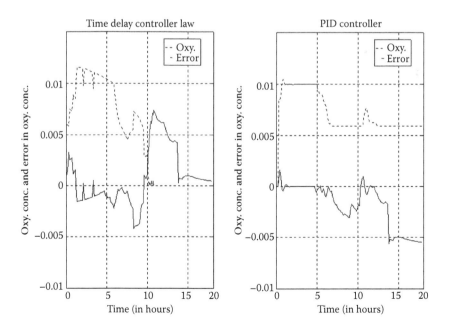

FIGURE 5.38 Comparison of time delay controller and PID for oxygen concentration control in op-mode 8.

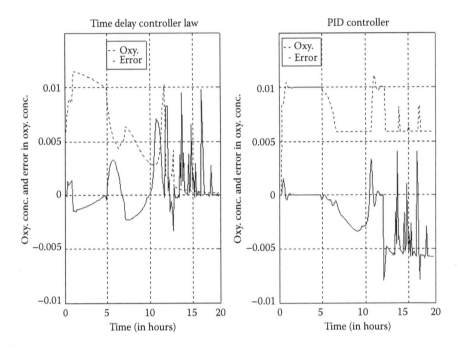

FIGURE 5.39 Comparison of time delay controller and PID for oxygen concentration control in op-mode 9.

It can be concluded that the time delay controller works better in the disturbed condition (as in the mixed mode, op-mode 9) than the PID controller. The reverse is the case when there is no disturbance. In a far-fetched way, this proves the utility of time delay controller in bioprocess system control where uncertainty in the model and disturbances could always affect the system.

The above simulation results explore the possibility of using TDC for bioprocesses. The simulation results show that it is suited for the purpose and computationally not very cumbersome. The observer design is also simplified because of the LTI reference model. Experimental verification of the time delay controller has been taken up in the laboratory. This is discussed in Chapter 6.

CHAPTER SIX

Experimentation on the bioreactor

A working diagram of the developed see-saw bioreactor with its instrumentation is shown in Figure 6.1. The operating procedure has already been discussed in Chapter 2. The mass transfer phenomenon of gaseous oxygen to the liquid medium of the bioreactor has also been explained in Chapter 2.

The instrumentation part of the bioreactor is explained in Figure 6.2.

6.1 Instrumentation

There are three transducers that can be sterilized attached to the bioreactor:

1. PT-100 RTD for temperature measurement (j in Figure 6.1)

2. Sensor for pH measurement (k in Figure 6.1)

3. Sensor for dissolved oxygen concentration measurement (l in Figure 6.1)

Temperature, pH, and dissolved oxygen concentration signals from these sensors are connected to the corresponding transmitters, which are installed very near to them. The sensor signals are amplified, filtered, and converted to proportional 4–20 mA current signals for transmission. Generally, this type of bioreactor is kept in a sterilized room and the process control computer, which may require human attention, is kept in a separate control room away from the bioreactor

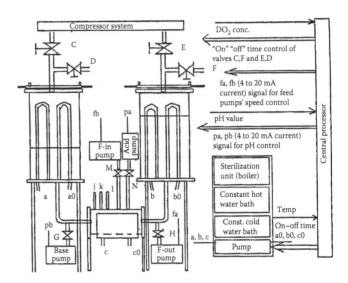

FIGURE 6.1 Schematic diagram of the bioreactor with its instrumentation.

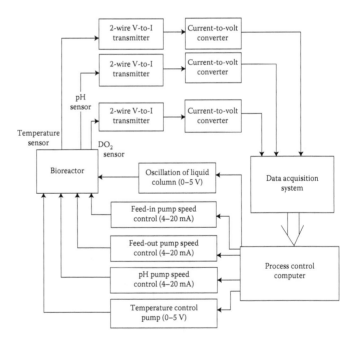

FIGURE 6.2 Block diagram of instrumentation and control of see-saw bioreactor.

system (consisting of the bioreactor, sensors and transmitters, compressor, and boiler for the purpose of sterilization). The transmitters use two wire transmission systems. However, the fast data acquisition system (PCL 208A card) accepts signals in voltage format. So, the current signal is again converted to a voltage signal. A special current-to-voltage converter with electrically isolated input and output is used. The signals go to the data acquisition system.

Temperature and pH signals are used to regulate the temperature and pH to preset values. Temperature control is achieved by adjusting the ON time of either of the constant-temperature (low- or high-temperature) water feed pumps so that the reference set point is reached. In the case of pH control, the speed of either of the two delivery pumps (acid or alkali) is controlled so that the pH set value is reached. This is discussed below in the section "Presentation of experimental results." The dissolved oxygen concentration signal is utilized for process control. This is the single output signal from which other state variables are estimated and the time delay control (TDC) law calculates the control action. The control action includes, among other things, speed of feed-in and feed-out pumps and the period of oscillation of the bioreactor. The programs meant for temperature and pH control and the control of the bioprocess are all written in MATLAB®. Temperature and pH controllers are on–off and proportional-type controllers and are operated serially. Process control is done separately. The environment controller and time delay controller work simultaneously.

6.2 Experimentation

Experimentation is carried out on a yeast culture. The aim of the experiment is to obtain a higher biomass concentration of yeast utilizing two different operating modes. The substrate composition used for the purpose is as follows:

Maltose → 3 g/L

Yeast extract → 3 g/L

Peptone → 5 g/L

Dextrose → 10 g/L

Experimentation was done in two operating modes: op-mode 5 (fed-batch mode) and op-mode 7 (continuous mode).

The total time of experimentation varied between 8 and 12 h. The initial concentrations of different states were as follows:

Initial biomass concentration → 0.5 g/L

Initial substrate concentration → 10 g/L

Initial dissolved oxygen concentration → 0.0033 g/L

Working volume of the bioreactor to start with → 8 L

The process model used for the purpose of control and state estimation is as follows:

$$\frac{d\,x_1(t)}{dt} = \frac{\mu_m\,x_1(t)x_2(t)x_3(t)}{(k_S + x_2(t))(k_C + x_3(t))}$$
$$- k_d x_1(t) + \frac{F_0(t)}{x_4(t)}x_{in} - \frac{F_1(t)}{x_4(t)}x_1(t) \qquad (6.1)$$

$$\frac{d\,x_2(t)}{dt} = -\frac{\mu_m\,x_1(t)x_2(t)x_3(t)}{Y(k_S + x_2(t))(k_C + x_3(t))}$$
$$- m_S x_1(t) + \frac{F_0(t)}{x_4(t)}S_{in} - \frac{F_1(t)}{x_4(t)}x_2(t) \qquad (6.2)$$

$$\frac{d\,x_3(t)}{dt} = -\frac{\mu_m\,x_1(t)x_2(t)x_3(t)}{Y_0(k_S + x_2(t))(k_C + x_3(t))}$$
$$- m_{S0} x_1(t) + \frac{F_0(t)}{x_4(t)}0_{in} - \frac{F_1(t)}{x_4(t)}x_3(t)$$
$$+ K_L a(0_2^* - x_3(t)) \qquad (6.3)$$

$$\frac{d\,x_4(t)}{dt} = F_0(t) - F_1(t) \qquad (6.4)$$

where
- $x_1(t)$, $x_2(t)$, and $x_3(t)$ represent the concentrations of the cell mass, substrate, and oxygen, respectively, in the liquid phase of the bioreactor in g/m^3.
- $x_4(t)$ denotes the working volume of the bioreactor in m^3. The time unit is measured in hours.
- F and F_1 represent the liquid feed rate and withdrawal rate using peristaltic pumps in the bioreactor, respectively (referring to Figure 3.1).
- x_{in}, S_{in}, and O_{in} are the influent cell mass, substrate, and oxygen concentrations, respectively.
- $K_L a$ denotes the oxygen mass transfer rate.
- K_S and K_C represent the saturation constants.

K_d, m_S, m_{S0}, Y, and Y_0 represent the biomass decay rate, maintenance coefficient with respect to carbon and oxygen source, yield coefficient with respect to carbon and oxygen source for cell mass growth, respectively.

O_2^* represents the saturation value of the oxygen concentration in the liquid medium of interest.

$\mu(x,t)$ represents the specific growth rate which can be assumed to have bounds as $0 \leq \mu \leq \mu_m$ for all x. μ_m is representing the maximal growth rate.

Considering Equations 6.1 through 6.4 as the process model, genetic algorithm was applied for op-modes 5 and 7 with an objective to achieve maximum biomass concentration (as discussed in Chapter 4).

Presentation of experimental results

After carrying out the experiment with this yeast strain using the process control software (discussed in Chapter 5) for feed control (rpm of feed-in and feed-out pumps and time period of oscillation of bioreactor liquid column for the control of mass transfer of gaseous oxygen to liquid medium) and environmental control (temperature and pH), the results are obtained. Cell mass and substrate concentrations are determined through off-line laboratory analysis.

During experimentation the dissolved oxygen concentration was the only measurable parameter. Other process variables (states) are estimated from the measured oxygen concentration using the observer.

Table 6.1 displays the *theoretical optimal results* (reference trajectory) and the corresponding *actual off-line measured*

Table 6.1 Comparison of experimental vs theoretical optimal results in op-mode 7 for verification of the controller

	For cell mass concentration			For substrate concentration		
Hour	Measured	Optimal	Error	Measured	Optimal	Error
0	0.333	0.5165	0.1832	10.2080	10.0827	−0.1253
1	0.7333	0.6549	−0.0785	10.2840	10.8032	0.5192
2	0.8667	0.8533	−0.0134	10.2480	11.4311	1.1831
3	0.9000	1.1118	0.2118	10.2080	11.8627	1.6547
4	1.000	1.4485	0.4485	10.1680	12.0799	1.9119
5	1.1333	1.7878	0.6545	8.8560	11.3641	2.5081
6	1.7667	2.2150	0.4483	6.8520	10.4213	3.5693
7.3	2.9000	3.0116	0.1116	5.4240	8.7853	3.3613

Table 6.2 Comparison of experimental vs theoretical optimal results in op-mode 7 for verification of the software sensor

	For cell mass concentration			For substrate concentration		
Hour	Measured	Optimal	Error	Measured	Optimal	Error
0	0.333	0.5168	0.1832	10.2080	9.9706	–0.2374
1	0.7333	0.6460	–0.0874	10.2840	10.3761	0.0921
2	0.8667	0.8426	–0.0241	10.2480	10.9993	0.7513
3	0.9000	1.0963	0.1963	10.2080	11.4717	1.2637
4	1.000	1.4203	0.4203	10.1680	11.7714	1.6034
5	1.1333	1.7643	0.6310	8.8560	11.2913	2.4353
6	1.7667	2.1710	0.4044	6.8520	10.3995	3.5475
7.3	2.9000	2.8584	–0.0416	5.4240	8.9784	3.5544

results for cell mass and substrate concentrations after application of the proposed time delay controller in op-mode 7.

Table 6.2 compares the on-line estimated results (the results from the observer) with the corresponding experimental results. Figures 6.3 through 6.12 give the results of different experimentations in op-mode 7. Similarly, Tables 6.3 and 6.4 repeat the same for op-mode 5, and Figures 6.13 through 6.22 give the results of experimentation.

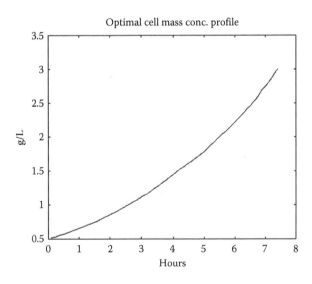

FIGURE 6.3 Optimal cell mass concentration in op-mode 7, that is, continuous mode.

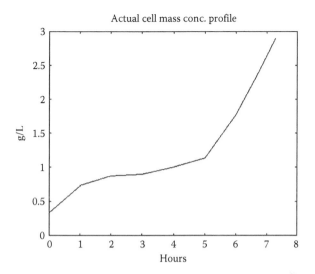

FIGURE 6.4 Measured cell mass concentration in op-mode 7, that is, continuous mode.

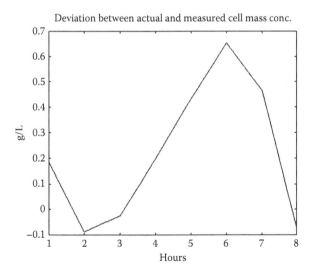

FIGURE 6.5 Deviation between actual and optimal cell mass concentrations in op-mode 7, that is, continuous mode.

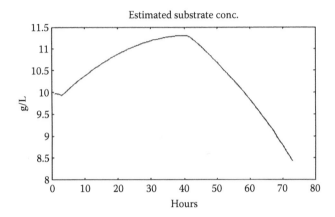

FIGURE 6.6 Optimal substrate concentration in op-mode 7, that is, continuous mode.

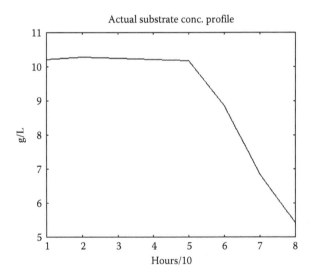

FIGURE 6.7 Measured substrate concentration in op-mode 7, that is, continuous mode.

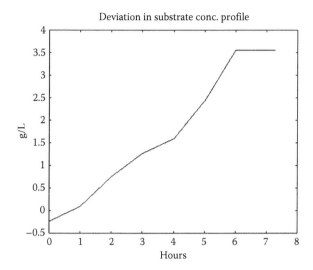

FIGURE 6.8 Deviation between actual and optimal substrate concentrations in op-mode 7, that is, continuous mode.

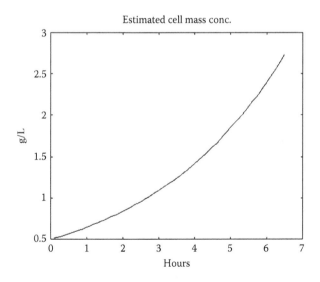

FIGURE 6.9 Estimated cell mass concentration in op-mode 7, that is, continuous mode.

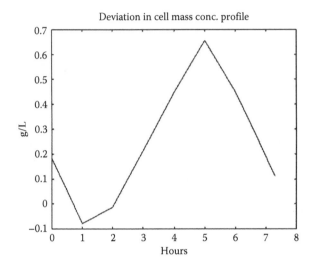

FIGURE 6.10 Deviation between actual and estimated cell mass concentrations in op-mode 7, that is, continuous mode.

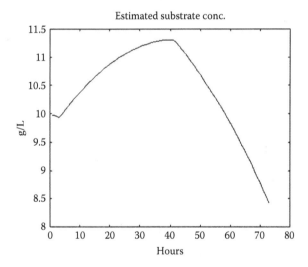

FIGURE 6.11 Estimated substrate concentration in op-mode 7, that is, continuous mode.

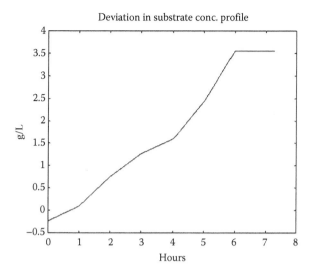

FIGURE 6.12   Deviation between actual and estimated substrate concentrations in op-mode 7, that is, continuous mode.

Table 6.3   Comparison of experimental vs theoretical optimal results in op-mode 5 for verification of the controller

	For cell mass concentration			For substrate concentration		
Hour	Measured	Optimal	Error	Measured	Optimal	Error
0	0.5122	0.5168	0.0046	10.0000	9.9724	−0.0276
1	0.5858	0.6554	0.0696	9.1964	10.1523	0.9558
2	0.7046	0.8535	0.1489	8.9881	10.2535	1.2654
3	0.7244	1.1112	0.3868	8.7165	10.2291	1.5125
4	0.9026	1.4466	0.5440	8.6905	10.0471	1.3566
5	2.7060	1.8830	−0.8230	8.5268	9.6672	1.1404
6	3.3660	2.4505	−0.7835	7.7976	9.0367	1.2390
6.9	3.7620	3.1056	−0.6564	7.3363	8.1981	0.8618

Table 6.4   Comparison of experimental vs theoretical optimal results in op-mode 5 for verification of the software sensor

Hour	For cell mass concentration			For substrate concentration		
	Measured	Optimal	Error	Measured	Optimal	Error
0	0.5122	0.5168	0.0046	10.0000	9.9706	−0.0294
1	0.5858	0.6449	0.0591	9.1964	9.9504	0.7540
2	0.7046	0.8390	0.1344	8.9881	9.8984	0.9103
3	0.7244	1.0913	0.3669	8.7165	9.7324	1.0159
4	0.9026	1.4192	0.5116	8.6905	9.4280	0.7375
5	2.7060	1.8456	−0.8604	8.5268	8.9487	0.4219
6	3.2340	2.3999	−0.8341	7.7976	8.2430	0.4454

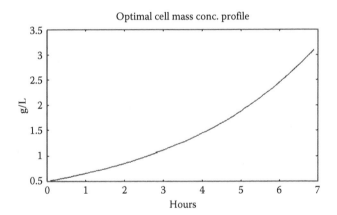

FIGURE 6.13   Optimal cell mass concentration in op-mode 5, that is, fed-batch mode.

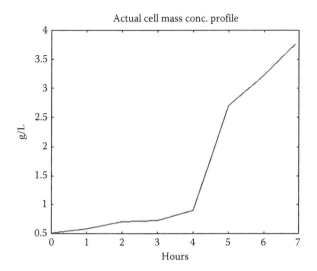

FIGURE 6.14 Measured cell mass concentration in op-mode 5, that is, fed-batch mode.

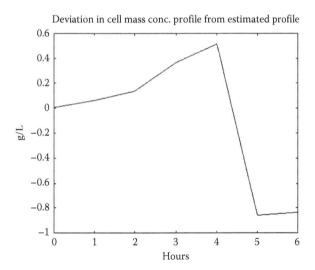

FIGURE 6.15 Deviation between actual and optimal cell mass concentrations in op-mode 5, that is, fed-batch mode.

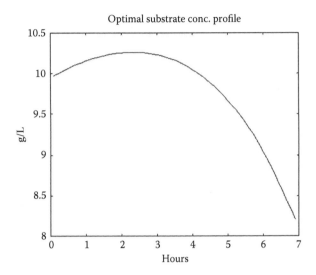

FIGURE 6.16 Optimal substrate concentration in op-mode 5, that is, fed-batch mode.

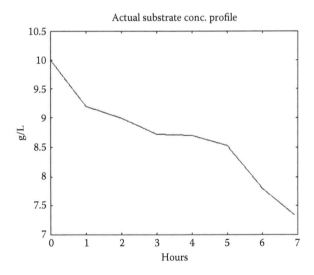

FIGURE 6.17 Measured substrate concentration in op-mode 5, that is, fed-batch mode.

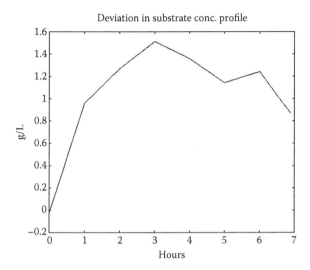

FIGURE 6.18 Deviation between actual and optimal substrate concentrations in op-mode 5, that is, fed-batch mode.

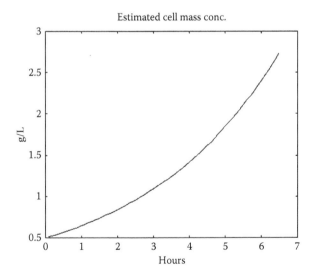

FIGURE 6.19 Estimated cell mass concentration in op-mode 5, that is, fed-batch mode.

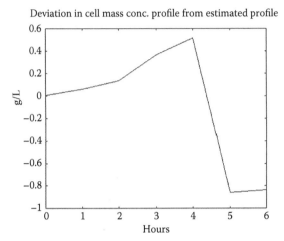

FIGURE 6.20 Deviation between actual and estimated cell mass concentrations in op-mode 5, that is, fed-batch mode.

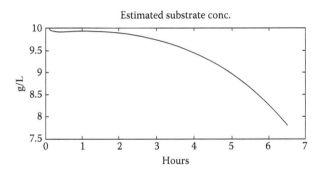

FIGURE 6.21 Estimated substrate concentration in op-mode 5, that is, fed-batch mode.

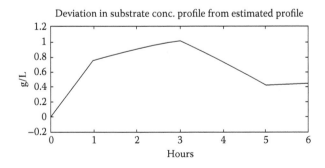

FIGURE 6.22 Deviation between actual and estimated substrate concentrations in op-mode 5, that is, fed-batch mode.

Figures 6.9 and 6.11 display the *estimated* profiles of the cell mass and substrate concentrations, respectively, in op-mode 7 (continuous mode). Similarly, Figures 6.19 and 6.21 display the same in op-mode 5 (fed-batch mode). Figures 6.10 and 6.12 show the error profile, that is, the difference between the experimental and estimated results with respect to the cell mass and substrate concentrations, respectively, in op-mode 7. Similarly, Figures 6.20 and 6.22 represent the error between the experimental and estimated results with respect to the cell mass and substrate concentrations, respectively, in op-mode 5.

Figure 6.23 displays the theoretically simulated control inputs to be applied in op-mode 7. Figure 6.24 displays the experimental control inputs applied in this operating mode. There was very little deviation as far as the revolutions per minute of the feed-out pump is concerned. Similarly, Figures 6.25 and 6.26 display the theoretical and actually applied control inputs in op-mode 5 (fed-batch mode).

Figures 6.3 and 6.13 show the optimal time profiles of cell mass concentration in op-modes 7 and 5, respectively.

Figures 6.4 and 6.14 show the measured time profiles of cell mass concentration in op-modes 7 and 5, respectively.

Comparison between experimental and theoretical results
Figures 6.5 and 6.15 show the deviations between the calculated optimal and experimental results with respect to cell mass

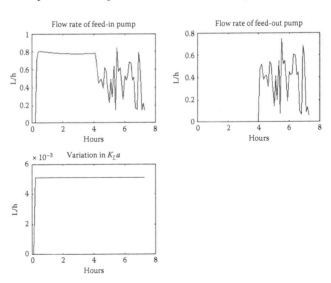

FIGURE 6.23 Time profiles of theoretical control inputs in op-mode 7, that is, continuous mode.

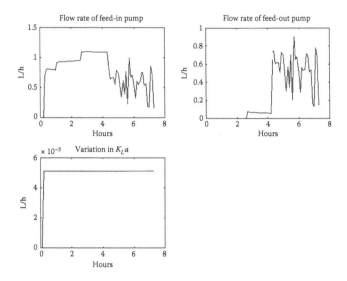

FIGURE 6.24 Time profiles of control inputs actually applied in op-mode 7, that is, continuous mode.

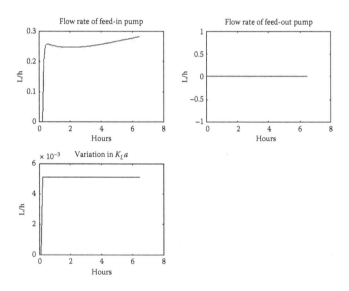

FIGURE 6.25 Time profiles of theoretical control inputs in op-mode 5, that is, fed-batch mode.

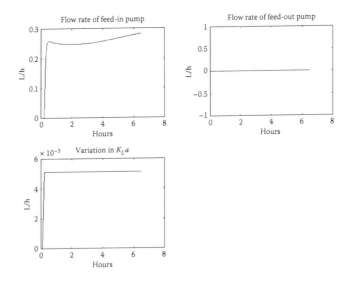

FIGURE 6.26 Time profiles of actual control inputs in op-mode 5, that is, fed-batch mode.

concentration in op-modes 7 and 5, respectively. Similarly, Figures 6.6 and 6.16 show the optimal time profiles of substrate concentration in op-modes 7 and 5, respectively.

Figures 6.7 and 6.17 show the measured time profiles of substrate concentration in op-modes 7 and 5, respectively.

Figures 6.8 and 6.18 show the deviations between the theoretical and experimental results with respect to substrate concentration in op-modes 7 and 5, respectively. From the above it could be inferred that the time delay controller works fairly well and establishes its validity in bioprocess control.

6.3 Summary

The deviations in the performance between theoretically predicted and experimental results are attributed to the following:

1. The mixing performance of this bioreactor is not to expectation. Alcohol is produced in pockets where the dissolved oxygen concentration value reaches very low. This inhibits growth. This factor was not incorporated in the system model.

2. The expression obtained for predicting K_La was fairly approximate. The same has been used in controller design.

3. As time passes, the viscosity of the bioreactor medium increases. This further restricts oxygen supply to the bioreactor system. This was also not taken care of in the system model.

4. The initial cell mass and oxygen concentration were not the same as the theoretical starting values.

5. The growth of cells (yeast) depends on inoculum age and other factors.

These are major reasons, apart from other unknown biological reasons. It can be inferred that the time delay controller worked satisfactorily in the case of yeast biomass production in the new prototype see-saw bioreactor [49]. The software sensor is also satisfactory. The validity of application of the time delay controller along with the observer system in bioprocess control is demonstrated. The same investigation may be carried out for other fermentation processes.

CHAPTER SEVEN

General conclusion and future scope of research

7.1 Overview of the work

In the present work, the main aim was to design and fabricate a see-saw bioreactor, suitable for the culturing of animal cell lines. This has been done successfully. A model was developed that approximately predicted the mass transfer of gaseous oxygen in the bioreactor medium.

The next objective was to integrate the bioreactor with a proper instrumentation system so that a state-of-the-art controller could be incorporated in the system for yield maximization. For this reason, temperature, pH, and dissolved oxygen probes were fitted to the bioreactor assembly along with suitable current transmitters. The signals from the transmitters were then connected to the computer through proper data acquisition arrangements.

For the bioprocess controller, first a generalized model with double-substrate limitation was framed. The bioprocess model considered is basically an unsegregated and unstructured model having four bioprocess state variables and four control inputs. The model considered is nonlinear, having time-varying parameters. In addition, the model has a lot of unmodeled dynamics. The only measurable bioprocess variable is the dissolved oxygen concentration. So the process demands a software sensor to estimate the unmeasurable bioprocess state variables. Time delay control (TDC) is used for the bioprocess. To generate reference trajectories for the bioprocess variables, genetic algorithm has been used.

The working of TDC was checked by simulation experiments in different operating modes. This was followed by experimentation on a fabricated prototype bioreactor. The control inputs calculated by applying the TDC were converted to current signals using a digital-to-analog converter (DAC) and voltage to current (V to I) converter. These current signals are used to regulate the speed of the respective peristaltic pumps for feed-in and feed-out flow. Samples were collected at 1 h intervals. The results of the software sensor and control inputs are plotted. At the end of the experiment, samples collected were analyzed in the laboratory to find the actual cell mass and substrate concentrations. These off-line analysis results were compared with the corresponding reference values to find the errors. The developed TDC law has been so verified. The analyzed experimental results were also compared with results of the software sensor (estimated values). Thus the utility of the observer has been checked.

The experimental work has been carried out in two different operating modes: 5 and 7. The results indicate that the error bounds are within 20%.

The environmental control, that is, the control of the temperature, and pH of the reactor medium to their predefined values were done using on–off and proportional controllers, respectively. Environmental controls remained operative for the whole fermentation period.

7.2 General conclusion

The bioreactor developed in the laboratory scale exhibited perfect airtightness. Sterilization was achieved as described in the appendix. The mass transfer model derived to predict the amount of dissolved oxygen concentration in the bioreactor medium with regard to oscillation of the liquid column with different time periods is shown in Figures 2.3 through 2.5. It can be concluded that the model requires further refinement. Presently, it describes the approximate oxygen mass transfer phenomenon [50]. There is also room for improvement in the design aspect of the fabricated bioreactor.

An expert simulation package BIPROSIM was developed using the bioprocess model. We propose that a bioreactor can be operated in nine operating modes (including the mixed mode). BIPROSIM serves the purpose of predicting the best operating mode for a particular fermentation process. BIPROSIM has been further improved using genetic algorithms. The same predicts how to maneuver control inputs so as to achieve maximum yield. This will help increase the yield in real processes. The resultant time profiles of the bioprocess variables will serve as reference trajectories for the bioprocess controller. Thus the simulation package helps save time and money, both of which are required to perform real-time experiments to investigate the same.

We have used TDC as a robust bioprocess controller along with a simple observer. TDC and the developed observer were examined by simulation for all the operating modes. The results were satisfactory. This was then applied to a real-time fermentation process with yeast. The aim was to achieve higher cell mass growth. The result of the application was good. The off-line experimental results (cell mass concentration, substrate concentration) differ from the corresponding optimal reference values within 20%.

It can generally be concluded that the results of the proposed controller and observer are encouraging. The performance could also be examined with other fermentation processes.

There is still work to be done to make the see-saw bioreactor system more attractive and a commercial proposition.

7.3 Future scope of research

1. The see-saw bioreactor developed is a laboratory prototype. Before commercialization, there is scope for further development in its design.

2. The bioreactor has been developed for animal cell line culture. Due to lack of facility, this could not be put to the same use. Experimentation for validation has been done on yeast. There remains scope for further experimentation, particularly on the culturing of animal cell lines.

3. Different types of bioprocess control strategies can be tested and verified in the setup.

4. Mass transfer of gaseous oxygen into liquid medium can be further enhanced by incorporating extended surfaces inside the reactor. Studies can also be carried out using packed beds (of porous ceramic beads) or other arrangements.

5. The model depicting the mass transfer phenomenon of oxygen from a gaseous to a liquid phase needs further refinement.

References

1. Saha, G., A. Barua, and S. Sinha, Development and modeling of a novel "see-saw" bioreactor, unpublished.

2. Holmberg, A. and J. Ranta, Procedures for parameter and state estimation of microbial growth process models, *Automatica*, vol. 18, no. 2, pp. 181–193, 1982.

3. Dochain, D., On-line parameter estimation, adaptive state estimation and adaptive control of fermentation process, PhD thesis, Universite Catholique de Louvain, Belgium, 1986.

4. Axelsson, J. P., Modelling and control of fermentation processes, PhD thesis, Department of Automatic Control, Lund Institute of Technology, 1989.

5. Seborg, D. E., T. F. Edgar, and D. A. Mellichamp, *Process Dynamics and Control*, Wiley, New York, 1989.

6. Astrom, K. J. and B. Wittenmark, *Adaptive Control*, Addison-Wesley, Reading, MA, 1988.

7. Driankav, H., *An Introduction to Fuzzy Control*, Narosa Publishing, India, 1993.

8. Czogala, E. and T. Rawlik, Modelling of a fuzzy controller with application to the control of biological processes, *Fuzzy Sets and Systems*, vol. 31, pp. 13–22, 1989.

9. Postlethwaite, B. E., A fuzzy state estimator for fed-batch fermentation, *Chemical Engineering Research & Design*, vol. 67, no. 3, pp. 267–272, 1989.

10. Tong, R. M., A retrospective view of fuzzy control systems, *Fuzzy Sets and Systems*, vol. 14, pp. 199–210, 1984.

11. Kosko, B., *Neural Networks and Fuzzy Systems*, Prentice-Hall, India, 1994.

12. Turunen, I. et al., Fuzzy modeling in biotechnology: Sucrose inversion, *Chemical Engineering Journal*, vol. 30, pp. B51–B60, 1985.

13. San, K. and G. N. Stephanopoulos, Studies on on-line bioreactor identification 2: Numerical and experimental results, *Biotechnology and Bioengineering*, vol. 26, pp. 1189–1197, 1984.

14. San, K. and G. N. Stephanopoulos, Studies on on-line bioreactor identification 4: Utilization of pH measurement for product estimation, *Biotechnology and Bioengineering*, vol. 26, pp. 1209–1218, 1984.
15. San, K. and G. N. Stephanopoulos, Studies on on-line bioreactor identification 1: Theory, *Biotechnology and Bioengineering*, vol. 26, pp. 1176–1188, 1984.
16. Saha, G., A. Barua, and S. Sinha, Dynamic optimization of bioprocesses using genetic algorithms, to be communicated.
17. Fredrickson, A. G., R. D. Magee, and H. M. Tsuchiya, Mathematical models for fermentation processes, *Advances in Applied Microbiology*, vol. 13, p. 419, 1979.
18. Brotherton, J. D. and P. C. Chau, Modeling analysis of an intercalated spiral alternate dead ended hollow fiber bioreactor, for mammalian cell cultures, *Biotechnology and Bioengineering*, vol. 35, pp. 375–394, 1990.
19. Park, S. and G. Stephanopoulos, Packed bed bioreactor with porous ceramic beads for animal cell culture, *Biotechnology and Bioengineering*, vol. 41, pp. 25–34, 1993.
20. Wang, N. S. and G. Stephanopoulos, Application of macroscopic balances to the identification of gross measurement error, *Biotechnology and Bioengineering*, vol. 25, pp. 2177–2208, 1983.
21. Henzler, H. J., Verfahrenstechnische auslegungsgrundlagen fur ruhrbehalter als fermenter, *Chemie Ingenieur Technik*, vol. 54, pp. 461–476, 1982.
22. Bird, R. O., W. E. Stewart, and E. N. Lightfoot, *Transport Phenomena*, Wiley, New York, 1960.
23. Ohno, H. and E. Nakanishi, Optimum operating mode for a class of fermentation, *Biotechnology and Bioengineering*, vol. 20, pp. 625–636, 1978.
24. Bastin, G. and D. Dochain, *On Line Estimation and Adaptive Control of Bioreactors*, Elsevier, Amsterdam, Netherlands, 1990.
25. Nielsen, J., Simulation of bioreactors, *Computers Chemical Engineering*, vol. 18, pp. 615–620, 1994.
26. Gebicke, K. W. and H. J. Johl, Application of modeling and simulation for optimization of a continuous fermentation process, *European Symposium on Computer Aided Process Engineering*, vol. 2, pp. 5177–5182, 1994.
27. Sinclair, C. G. and D. N. Ryder, Models for continuous culture of micro-organisms under both carbon and oxygen limiting conditions, *Biotechnology and Bioengineering*, vol. 17, pp. 375–398, 1975.
28. Williams, F. M., A model for cell growth dynamics, *Journal of Theoretical Biology*, vol. 15, no. 2, pp. 190–207, 1967.
29. Svrcek, W. Y., R. E. Elliott, and J. E. Zajic, The extended Kalman filter applied to a continuous culture model, *Biotechnology and Bioengineering*, vol. 16, no. 6, pp. 827–846, 1974.

30. Chattaway, T. and G. N. Stephanopoulos, An adaptive state estimator for detecting contaminants in bioreactors, *Biotechnology and Bioengineering*, vol. 34, pp. 647–659, 1989.
31. Saha, G., A. Barua, and S. Sinha, BIPROSIM: A generic package for bioprocess simulation, communicated to *Computers and Chemical Engineering*.
32. Goldberg, D. E., *Genetic Algorithms in Search, Optimization, and Machine Learning*, Addison-Wesley, Boston, 1989.
33. Nielsen, J., K. Nikolajsen, and J. Villadsen, Structured modelling of a microbial system, *Biotechnology and Bioengineering*, vol. 38, pp. 11–23, 1991.
34. Bader, F. G., Analysis of double substrate limited growth, *Biotechnology and Bioengineering*, vol. 20, pp. 183–202, 1978.
35. Proll, T. and N. M. Karim, Non-linear control of bioreactor model using exact and I/O linearisation, *International Journal of Control*, vol. 60, pp. 499–519, 1994.
36. Luttmann, R., G. Bitzer, and J. Hartkopf, Development of control strategies for high cell density cultivations, *Mathematics and Computers in Simulation*, vol. 37, pp. 153–164, 1994.
37. Chen, L., G. Bastin, and V. V. Breusegem, A case-study of adaptive non-linear regulation of fed-batch biological reactors, *Automatica*, vol. 31, pp. 55–65, 1995.
38. Johnson, A., The control of fed-batch fermentation processes: A survey, *Automatica*, vol. 23, pp. 691–705, 1987.
39. Ohno, H., E. Nakanishi, and T. Takamatsu, Optimal control of semibatch fermentation, *Biotechnology and Bioengineering*, vol. 18, pp. 847–864, 1976.
40. Chen, L., G. Bastin, and V. V. Breusegem, A case study of non-linear regulation of a fed-batch biological reactor, *Automatica*, vol. 31, pp. 55–65, 1995.
41. Youcef-Toumi, K. and O. Ito, A time delay controller for systems with unknown dynamics, *Journal of Dynamic Systems, Measurement, and Control*, vol. 112, pp. 133–142, 1990.
42. Youcef-Toumi, K. and S. Reddy, Analysis of linear time-invariant systems with time delay, in *Proceedings of American Control Conference*, Chicago, June 24–26, 1992, pp. 1940–1944.
43. Chang, P. H., A model reference observer for time delay control and its application to robot trajectory control, *IEEE Transactions on Control System Technology*, vol. 4, pp. 2–10, 1996.
44. Chang, P. H. and J. W. Lee, An observer design for time delay control and its application to DC servo motor, *Control Engineering Practice*, vol. 2, pp. 263–270, 1994.
45. Shi, Z. and K. Shimizu, Neuro-fuzzy control of bioreactor systems with pattern recognition, *Journal of Fermentation and Bioengineering*, vol. 74, pp. 39–45, 1992.
46. Bailey, J. E. and D. F. Ollis, *Biochemical Engineering Fundamentals*, McGraw Hill, New York, 1986.

47. Papageorgakopoulou, H. and W. J. Maier, A new modeling technique and computer simulation of bacterial growth, *Biotechnology and Bioengineering*, vol. 26, pp. 275–284, 1984.
48. Ogata, K., *Designing Linear Control Systems with MATLAB*, Prentice-Hall, Englewood Cliffs, NJ, 1994.
49. See-saw bioreactor, patent no. 199363, June 2006, India.
50. Saha, G., A. Barua, S. Sinha, B. C. Bhattacharya, and S. Ray, Modelling of the oxygen transfer characteristics of a "see-saw" bioreactor, *Chemical Engineering and Technology*, vol. 24, no. 1, pp. 97–101, 2001.

Appendix A:
Environmental control and sterilization of the bioreactor

A.1 Environmental control

A very important aspect of a bioprocess control is the environmental control inside the bioreactor. The process model discussed so far assumes that the environment is congenial for the bioreaction to continue. It is possible to incorporate congenital environment in the process model by studying the effect of the environment on μ (specific growth rate expression). This requires extensive experimentation for a specific process. Once this is done, environmental variables, specifically temperature and pH, will act as process control inputs. Thus the process controller would also control the temperature and pH in addition to F_0, F_1, F_4, and $K_L a$.

In the experimental setup, temperature and pH have not been treated as control inputs. Instead, they have been considered environmental variables. For any particular fermentation process, the favorable values of temperature and pH are investigated by actual experimentation. These particular values of temperature and pH are maintained (within a bound) at the corresponding set points during the whole fermentation period.

The controllers used for this purpose are the on–off type.

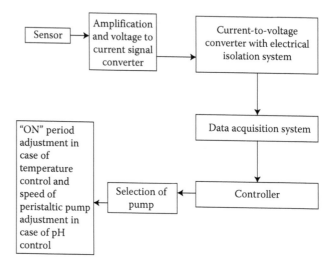

FIGURE A.1   Block diagram of the environmental controller.

Design of the temperature controller

The schematic structure for temperature control arrangement is shown in Figure A.1, and the actual arrangement is shown in Figure 6.1 (through the tubes indicated by a and a0, b and b0, and c and c0). As shown in Figure A.2, the temperature of the fluid inside the bioreactor is to be maintained at a preset temperature. This is done by flowing water at a specific temperature through the immersed U-tubes. The heat is either given to or carried away by the flowing water until the temperature of the fluid reaches the preset value.

The following assumptions are made:

1. Water at a fixed temperature flows inside the U-tubes at a high speed such that the temperature gradient at the inlet and the outlet is small.

2. Mixing of liquid inside the reactor is perfect.

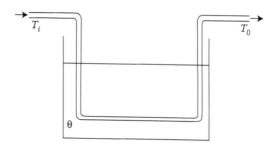

FIGURE A.2   Variables in the temperature controller.

The schematic diagram of the controller is shown in Figure A.3. The temperature controller algorithm is at below:

1. Given the initial and final temperatures of the bioreactor, the time required for the bioreactor temperature to reach the final temperature is calculated (t) from the mathematical model.

2. At time $t = (t/2)$, the temperature attained by the bioreactor liquid is calculated.

3. The outlet water temperature of the tube is also calculated from the mathematical model.

4. The actual temperature of the reactor is measured.

5. If (calculated theoretical temperature – actual measured temperature) differs by more than a predefined value, the heat transfer coefficient of the steel tubes is revalued.

6. For the next iteration,

   a. The water inlet temperature through the tube is replaced by the average of the inlet and outlet temperatures.

   b. The new value of the heat transfer coefficient is used.

   c. The initial reactor temperature is replaced by the measured one.

7. Steps 1–5 are repeated until the process temperature reads the predefined set temperature within the error bound.

There are two constant temperature baths (one hot, another cold) along with two pumps. If the measured temperature error, that is, the difference between the set point value and the measured value, is negative, then the cold water pump is energized and this will remain ON until the process temperature reaches the set value. If the error is positive, the hot water pump is energized to do the same.

Temperature measurement is done using a platinum PT-100 resistance temperature detector (RTD) with signal conditioning circuitry. The current signal is converted to a voltage signal before being fed to the data acquisition system.

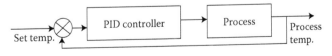

FIGURE A.3 Schematic diagram of the temperature controller.

The data acquisition system consists of a PCL 208A card with eight channel differential input ports and two channel digital-to-analog output ports. Data acquisition is carried out at an interval of every 6 min.

For the mathematical model and calculation of the different parameters in the above algorithm, let

θ_0 = initial temperature of the bioreactor liquid
θ_f = final temperature of the bioreactor liquid
T_i = inlet water temperature of the heat exchanger steel tube
T_0 = outlet water temperature of the heat exchanger liquid
$T'' = \dfrac{T_i + T_o}{2}$ = average inlet temperature
M = mass of bioreactor liquid
S = specific heat of water = 4180 J/kg/K
K = heat transfer coefficient for the steel tube
d = thickness of the steel tube
t_f = time to reach set temperature

From the heat balance equation,

$$-MS\frac{d\theta}{dt} = \frac{KA(\theta - T_i)}{d}$$

or

$$\frac{d\theta}{(\theta - T_i)} = \frac{KA}{d} \times \frac{1}{MS} \times dt$$

Integrating,

$$\int_{\theta_0}^{\theta_f} \frac{d\theta}{(\theta - T_i)} = \int_0^{t_f} -\frac{KA}{d} \times \frac{1}{MS} \times dt$$

or

$$\ln(\theta - T_i)\Big|_{\theta_0}^{\theta_f} = -\frac{KA}{d} \times \frac{1}{MS} \times t \Big|_0^{\theta_f}$$

or

$$\ln\frac{(\theta_f - T_i)}{(\theta_0 - T_i)} = -\frac{KA}{d} \times \frac{t_f}{MS}$$

Therefore,

$$t_f = -\frac{MSd}{KA} \ln \frac{(\theta_f - T_i)}{(\theta_0 - T_i)} \tag{A.1}$$

and

$$K = -\frac{MSd}{At_f} \ln \frac{(\theta_f - T_i)}{(\theta_0 - T_i)} \tag{A.2}$$

or

$$\frac{(\theta_f - T_i)}{(\theta_0 - T_i)} = e^{-\frac{KAt_f}{MSd}}$$

so that

$$\theta_f = \theta_0 e^{-\frac{KAt_f}{MSd}} + T_i \left(1 - e^{-\frac{KAt_f}{MSd}} \right) \tag{A.3}$$

The temperature of the water at the outlet of the heat exchanger tube is calculated as below. Let

$$\Delta T = T_0 - T_i$$

$$MS\Delta T = \frac{KA(\theta_f - T_i)}{d}$$

or

$$\Delta T = \frac{KA}{MSd} \times (\theta_f - T_i)$$

From Equation A.3 we have

$$\Delta T = \frac{KA}{MSd} \times \left(\theta_f e^{-\frac{KAt_f}{MSd}} - T_i e^{-\frac{KAt_f}{MSd}} \right)$$

or

$$T_0 = T_i + \frac{KA}{MSd} \times e^{-\frac{KAt_f}{MSd}} (\theta_f - T_i) \tag{A.4}$$

Now we calculate T_0 which is required in the calculation of

$$T'' = \frac{T_i + T_0}{2} \tag{A.5}$$

1. Equation A.1 gives the theoretical estimate for time for the bioreactor liquid to reach θ_f from θ_0.
2. At $t = (t_f/2)$, temperature of the bioreactor liquid is calculated using Equation A.3.
3. The temperature of the outlet water of the heat exchanger tube is calculated using Equation A.4.
4. If the measured value of the temperature of the bioreactor liquid is (T_m) and if $T_{th} \neq T_m$, then K is revalued using Equation A.2.
5. In the next iteration, put $\theta_0 = T_m$ and $T_i = T''$ as in Equation A.5, and K is revalued and the time required for the medium to reach θ_f is calculated.
6. The above-mentioned steps are repeated with new values of θ_i, T_i, and K until the process temperature reaches the set value.

Design of pH controller

To control the pH value of the bioreactor to a predefined set of values, two peristaltic pumps are used. One of the pumps is an acid delivery pump, and the other delivers alkali. The pH controller is a proportional-type controller. The input to the controller is the measured pH value of the bioreactor and the set value. The controlled variable is the speed (in rpm) of the peristaltic pump. The proportional controller output is the 4–20 mA current signal from the digital-to-analog converter (DAC). The TDC law is written and executed using MATLAB®.

A flowchart of the pH controller is shown in Figure A.4, where brephac, brephba, brepha7, brphacs, arephac, arephba, arephb7, and arphbas are different algorithms taking care of different situations in the pH control. However, two basic algorithms have been used. Algorithm A.1 discusses the control action when the solution is to be made acidic. Algorithm A.2 discusses the control action when the solution is to be made alkaline.

Algorithm A.1

1. Let the pH value of the standard acid solution be pH_s.
2. If the pH of the solution is x, then $[H^+] = 10^{-x}$.

3. The molar concentration of the given acidic solution (C_A) is given by $([H^+]^2 - 10^{-14}) / [H^+]_A$.

4. Let the reference pH be pH_r or $[H^+]_r = 10^{-pH_r}$.
   Thus the molar concentration of the reference solution (C_{Ar}) is $([H^+]_r^2 - 10^{-14}) / [H^+]_r$.

5. Similarly, the molar concentration of standard acidic solution C_{As} becomes $([H^+]_{As}^2 - 10^{-14}) / [H^+]_{As}$, where $[H^+]_{As} = 10^{-pH_s}$.

6. If V volume of standard acidic solution is needed to convert M volume of acidic solution of molar concentration C_A to an acidic solution of molar concentration C_{Ar} (set value), then V is calculated as $abs(M(C_A - C_{Ar})) = VC_{As}$ or

$$V = \frac{abs(M(C_A - C_{Ar}))}{C_{As}}$$

7. If this V volume of acid is dropped in the solution in time $t = 30$ s, then the flow rate is calculated as $f_1 = (v/t)$, and the corresponding rpm of the peristaltic pump is calculated as $f = (10/9)f_1$.

8. If $f >$ maximum available rpm of the pump, then set $f = f_{max}$ and recalculate
   $t = (v/f_1)$, and the corresponding flow rate f_l is calculated as $f_{max} = (10/9) \times f_1$.

9. The corresponding current signal (within 4–20 mA) from the DAC card to be sent to the peristaltic pump to drive it at the required rpm is $i_s = [(f \times (16/89.5)) + 4]$, where 89.5 rpm is the maximum possible speed of the peristaltic pump.

10. This signal is allowed to drive the pump up to time $t = (t/2)$.

11. After that, the pump is stopped by forcing a current signal $i_s = 4$ mA to the pump.

12. Mixing is carried out for 65 s.

13. The theoretical pH value (t_{pH}) is calculated as below:
    Adding $V_1 = (V/2)$ volume of acidic solution to the medium, the theoretical molar concentration C_n is calculated as

$$MC_A + V_1 C_{As} = MC_n$$

or

$$C_n = C_A + \frac{V_1}{M} C_{As}$$

And the theoretical pH (t_{pH}) is calculated as

$$[H^+] = C_n + \frac{10^{-14}}{[H^+]}$$

or

$$[H^+]^2 = C_n[H^+] + 10^{-14}$$

or

$$[H^+]^2 - C_n[H^+] - 10^{-14} = 0$$

So

$$[H^+] = \frac{C_n \pm \sqrt{C_n^2 + 4 \times 10^{-14}}}{2}$$

Select the nonzero, nonnegative value of $[H^+]$ whose ($-\log([H^+])$) value is close to pH_r, so that $t_{pH} = -\log_{10}([H^+])$.

14. Measure the actual pH of the solution (pH_m).
15. If $t_{pH} \neq$ measured $pH(pH_m)$, replace $[H^+]_A$ by 10^{-pH_m} for the next iteration.
16. Repeat steps 1–15 until $(pH - pH_r) \geq 0.05$.
17. Stop.

Algorithm A.2

(Basic/neutral solution to be made more basic)

1. Let the value of pH of the standard alkali solution be $pH = pH_s a$.
2. If the pH of the solution is xa, then $[H^+]_A = 10^{-(14-xa)}$.
3. The molar concentration of the basic solution becomes $C_b = \dfrac{([H^+]_A^2 - 10^{-14})}{[H^+]_A}$.
4. Let the reference pH value be $pH_r a$; then $[H^+]_r = 10^{-(14-pH_r a)}$. So the molar concentration of alkali of the reference pH_r valued solution becomes $C_{br} = \dfrac{([H^+]_r^2 - 10^{-14})}{[H^+]_r}$.

5. Similarly, the molar concentration of the standard alkali solution becomes $C_{bs} = \dfrac{([H^+]_s^2 - 10^{-14})}{[H^+]_s}$, where $[H^+]_s = 10^{-(14-pH_s a)}$.

Steps 6–17 are logically and numerically the same as Algorithm A.1 with the following deviation in step 13:

Select the nonnegative, nonzero value of $[H^+]$, whose $(abs(14 - abs(-\log_{10}([H^+]))) - pH_r a)$ value is minimum.

So $t_{pH} = (14 - abs(-\log_{10}([H^+])))$.

Algorithm arephac

This is meant for making an acidic solution more acidic.

Algorithm arephba

This is meant for making an acidic solution less acidic by pouring standard alkaline solution into it.

Algorithm arephb7

This is meant for making an acidic solution neutral by pouring alkaline solution in it (i.e., $pH_r = 7$).

Algorithm arpbhas

This is meant for making a neutral solution alkaline up to a certain reference pH value (pH_r).

Algorithm brephba

This is meant for making an alkaline solution more basic by adding standard alkaline solution to it.

Algorithm brephac

This is meant for making alkaline solution less basic by adding standard acidic solution to it.

Algorithm brepha7

This is meant for making alkaline solution neutral by adding standard acidic solution to it.

Algorithm brphacs

This will make neutral solution acidic by adding standard acidic solution to it.

As seen from the main flowchart (Figure A.4), if an acidic solution is to be made basic or vice versa, it must to go through the neutral. After that, the neutral solution is made alkaline or acidic.

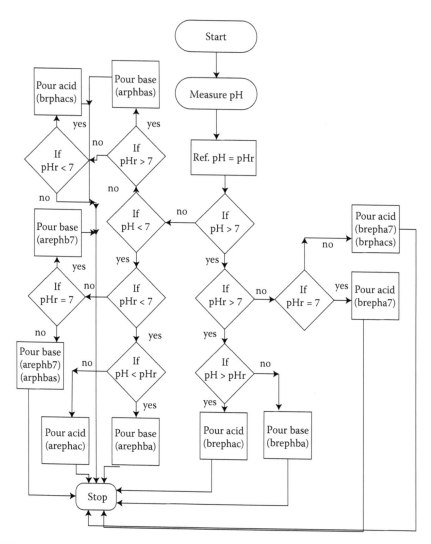

FIGURE A.4 Flowchart of pH control.

A.2 **Sterilization**

Sterilization is achieved by sending steam at 125°C to pass through ports a and a0, b and b0, and c and c0, as shown in Figure 6.1, which is meant for cooling and heating purposes of the bioreactor during normal operation. The temperature of the bioreactor medium is raised and kept at 121°C for 20 min. The specification of the boiler is as follows:

Wattage = 5 kW

Capacity = 3 L

Steam pressure = 2.1 bar

Fuel = heated electrically by three-phase 440 V AC supply

Index